D0629139

NUCLEAR POWER TRANSFORMATION

NUCLEAR POWER TRANSFORMATION

Joseph P. Tomain

Indiana University Press

BLOOMINGTON AND INDIANAPOLIS

Manufactured in the United States of America

Library of Congress Cataloging-in-Publication Data

Tomain, Joseph P., 1948–
Nuclear power transformation.

Includes index.
1. Nuclear industry—Law and legislation—
United States. 2. Nuclear energy—Government policy—
United States. 3. Nuclear industry—Government policy—
United States. 4. Nuclear energy—Government policy—
United States. I. Title.
KF2138.T66 1987 346.7304'6724 86-45397
ISBN 0-253-34110-8 347.306467924

1 2 3 4 5 91 90 89 88 87

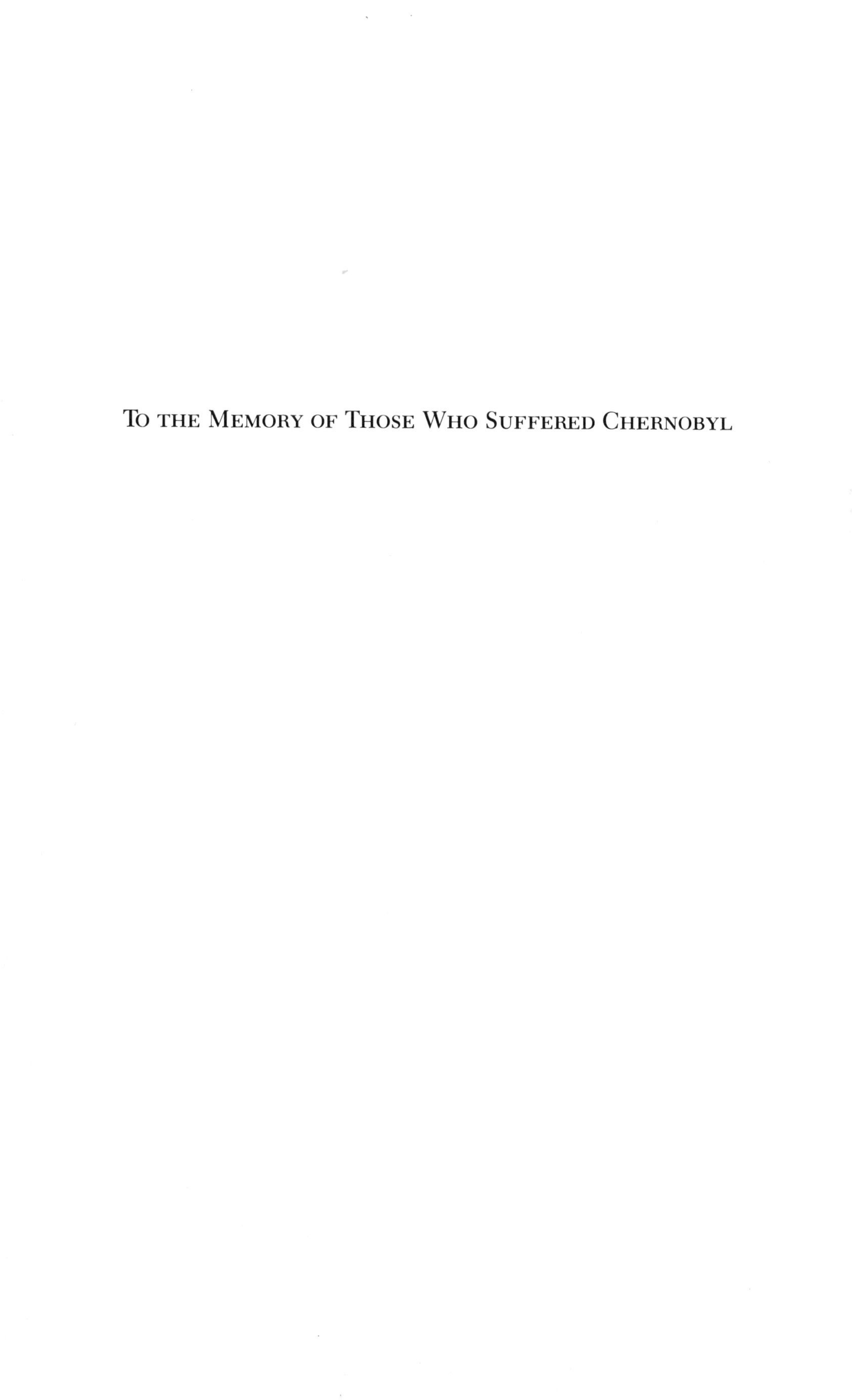

To the Memory of Those Who Suffered Chernobyl

Contents

PREFACE

This book can be subtitled "From Three Mile Island to Chernobyl." During this period, the regulation of the commercial nuclear power industry in the United States experienced a radical transition caused by dramatic changes in nuclear markets and nuclear politics. Unfortunately, nuclear power, once the hope and envy of energy suppliers, has turned out to be a costly mistake. Now, and for the foreseeable future, there will be no more investment in new nuclear plants. Instead, nuclear regulation will be occupied with existing and nearly on-line plants. Furthermore, unanticipated costs of nuclear power such as costs attributable to evacuation, plant decommissioning, and waste disposal are now being identified and allocated during this transitional period. The consequence of the transition is clear: There will be no more nuclear power plants constructed in the United States until costs are lowered and industry safety claims receive wide public acceptance. These are assertions about the economics and politics of nuclear power and they are the twin conditions for a resurgence of nuclear power.

The nuclear regulatory transition is not discrete in the sense that a conscious change in government policy was announced and implemented. Instead, the change is largely unconscious and is taking place gradually. The transition can be dated from March 28, 1979, the date of the incident at Three Mile Island, to April 26, 1986, the date of Chernobyl. TMI and Chernobyl serve as more than convenient mileposts in the history of nuclear power. TMI made the United States aware of unforeseen costs, just as Chernobyl made the world aware of unforeseen risks. These accidents are stark reminders of the complexities, risks, and costs of government-sponsored and regulated enterprises. Further, and more important, the events which occurred between these dates have significantly changed the direction of the nuclear and electric industries. *Nuclear Power Transformation* is about the economic, political, and legal dimensions of the transition.

This book is the end result of a collaborative effort. It began as a project in my Energy Law and Policy seminar in 1984 at the University of Cincinnati College of Law. A group of students and I were interested in examining the problems of nuclear plant cancellations and decided that we would each examine an aspect of the problem and write a chapter ultimately to be published as a book. As the project expanded and the students graduated, I altered the focus and theme of the book. Still, this book would not have been written without the help of several persons who deserve special thanks and recognition. Thomas Gabelman and James Jorling greatly contributed to the case studies in chapter two and to the history of nuclear regulation in chapter one. Lois Zettler and Jo Jones Riser developed the background for the discussion of public participation in chapter three. Without the facts uncovered by the diligent research of these four contributors, there would be no story. Rebekah Bell Estes tracked down the several state public utility commission cases in chapter five, and Lynn Schumacher and Michael Norse helped provide the discussion of government liability in chapter six. Finally, special thanks is due Constance Dowd Burton for her many contributions to the final form of this book. As my research assistant, she helped edit the manuscript and find obscure references. More important, through our many conversations and rereadings, she helped shape the final product.

The University of Cincinnati College of Law assisted this project with two summer grants, and provided the expert word processing help of Charlene Carpenter. My sincere appreciation to all who helped bring this work through to completion.

NUCLEAR POWER TRANSFORMATION

Introduction

A few seconds after 4 o'clock on the morning of March 28, 1979, pumps supplying feedwater to steam generators in the containment building of General Public Utility's Unit No. 2 near Harrisburg, Pennsylvania, closed down.[1] Automatically, emergency feedwater pumps kicked on, but a closed valve in each line prevented water from reaching the generators. The closed valves were not noticed by plant operators. As a result, another critical valve, the PORV, thought by plant operators to have closed after thirteen seconds, stuck open, sending critically needed coolant to the containment building floor. As the steam generators boiled dry, the reactor coolant heated and expanded. Two large pumps automatically began pouring coolant into the reactor chamber while pressure dropped. As a result of even more operator errors, known anomalously as common-mode failures,[2] these pumping systems, designated to send cooling water into the reactor vessel to reduce pressure and heat, were manually shut down. For critical hours, as water boiled into steam, the reactor failed to cool and began to disintegrate. The fuel rods crumbled, and gases within the rods escaped into the coolant water. After two hours and twenty-two minutes, a blocked valve was closed, stopping the flow of over 32,000 gallons of contaminated coolant into the containment building. Then thousands of gallons of deadly radioactive water were negligently pumped into an adjoining building. These events and those that followed are commonly referred to as the accident at Three Mile Island. TMI, a milestone in the history of commercial nuclear power, marks the end of its developmental period and the beginning of its transformation.[3]

The transformation is decisive. The once-unified pronuclear energy policy existing from the conclusion of World War II to 1979 has been shattered.

Currently, new commercial nuclear development has ended, and no consensus exists on what shape future policy will take. Before any recognizable policy is formed and implemented, political and economic interests influencing policy developments must work their way through a complex of decision-making processes. This book, about the transitional period of nuclear regulation, describes how and why the pronuclear preference disintegrated, sketches the contours of future policy, and argues that the new policy must confront institutional biases established during the developmental years of nuclear power. Briefly, the conclusion is that more responsive, democratic, and participatory decision-making processes are necessary before future nuclear policy achieves legitimacy.

Prior to TMI, the popular conception of a nuclear catastrophe was a core meltdown. The meltdown, colloquially referred to as the China Syndrome, is a nightmarish phenomenon in which the molten reactor core melts through thousands of tons of concrete and steel encasing the fuel rods and burns its way into the ground, emitting massive amounts of radioactive gas on its way to contaminating underground water tables. The radioactivity released into the atmosphere and the water system is predicted to cause over three thousand prompt fatalities, tens of thousands of illnesses, latent cancer fatalities, and genetic defects, and economic losses of $14 billion.[4] According to economist Daniel Ford, former executive director of the Union of Concerned Scientists, TMI was within thirty to sixty minutes of a core meltdown.[5]

As frightening as the vision of a health and safety holocaust is, the irony of TMI is the change in focus away from the radioactive to the financial consequences of a nuclear incident. Even at the writing of this book, the TMI story has not ended. There are two electric generating plants on the island. Because of the accident, both Units No. 1 and No. 2 were shut down. Unit No. 1 was permitted to reopen in 1985, and Unit No. 2 remains closed and contaminated. The clean-up costs of the TMI accident run over $1 billion,[6] multiples of previous estimates for partial plant decommissioning.[7] Costs continue to mount for clean-up, plant decommissioning, purchase of replacement power, and a welter of associated litigation,[8] including claims against the utility from those who suffered psychological damages. TMI is symbolic of nuclear plant cancellations, conversions, and delays costing tens to hundreds of billions of dollars. Collectively, the events surrounding TMI and its aftermath signal the beginning of an important new era for the commercial nuclear power industry and for its regulation by government. This era is captured by the question, Who pays?

The nuclear power industry has been brought to a halt primarily by market forces that have policy and regulatory reverberations. Commercial nuclear power, once believed to be the bright and shining hope for our energy future, has been stalled. Projections concerning the expansion of nuclear power plants have been revised continuously downward. In 1960, the government estimated 1,500 plants by the year 2000. The projections dwindled by the mid-1970s, when only 400 plants were planned. In 1981, only 78 additional plants were forecast.[9] The current prognosis is that in addition to the 77 nuclear plants already on line and generating electricity, about 50 more will be added, even though some 163 plants had been planned.[10] That no new nuclear power plant has been ordered since 1978 demonstrates the lack of faith that utility managers and private-sector investors have in the industry. The remaining 113 plants under construction are being either canceled or converted to burning fossil fuels, especially coal. In addition, most of the plants are experiencing undreamed-of postponements. The cancellations, conversions, and delays signify the abandonment of faith in commercial nuclear power. Nuclear abandonment may be, and most likely is, only temporary.[11] Nevertheless, because of the magnitude of the abandonment costs, society's response, particularly that of the regulatory establishment, affects future nuclear policy. The response also has an impact upon the relationship of government and industry, energy policy and politics, and law and legal institutions.

The country's overall energy planning needs an abandonment costs-allocation policy for the costs attributable to mistaken nuclear decisions. The policy, to be administered by various legal institutions, must recognize market signals and needs and must be responsive to changing political demands. Briefly, the market requires that costs be spread as efficiently as possible, which is best accomplished through uniform centralized decision making. The political reality is that nuclear power decision making is being rapidly decentralized at federal and state levels, and nuclear policy is severely fragmented as a consequence. Unfortunately, remnants of the past promotional policy are contained in the legal system now handling an unanticipated set of problems. The legal system was influenced by a promotional policy based on political and market factors no longer operating. Nevertheless, past policy shapes current legal decisions. This divergence between past policies and current needs and the conflict between politics and markets exemplify a major transition in nuclear regulation.

To fully describe and evaluate the new emerging regulatory structure, chapter one places nuclear regulation in its historical and economic contexts. Major legislative and judicial pronouncements will be discussed. Then, the interested parties are identified through case studies in chapter two. Chapter three continues the case discussion, with special emphasis on public participation. After the dramatis personae are identified, the market and the financial situation of the electric industry and of nuclear utilities are presented in chapter four. The nuclear market is such that billions of dollars have been mistakenly invested in construction of plants that produce no electricity and threaten to bankrupt utilities. In order to protect the electric industry, state and federal regulators are making decisions that accommodate the financial needs of utilities. These accommodations are explained in chapter five. In chapter six, the current system of legal liability rules is analyzed in order to suggest a direction for regulatory reform. Chapter seven concludes with an analysis of the necessary framework for future regulatory policy and institutional redesign.

Commercial nuclear power and its regulation move much by their own momentum. Government and industry invested heavily in nuclear power because it promised so much. Their faith was such that a healthy energy future would be assured by choosing the nuclear option. In the rush to meet the future, both government and industry created a regulatory structure promoting nuclear power without either party assuming concomitant responsibilities for having made the choice. Safety, environmental, and financial risks were passed from government to consumers and taxpayers, while industry insulated itself from those liabilities by passing risks to shareholders. Such was the faith in nuclear power. Faith has led to disillusionment, and disillusionment has led to a reconsideration of the place of nuclear power in our energy plan.

The narrowest focus, or core, of the book's discussion deals with how to spread the costs associated with "bad decisions" regarding nuclear power plants. *Bad* is as descriptive as it is evaluative. Nuclear plant construction costs, for example, have been outrageously uneconomical. Costs of delay, conversion, and cancellation are all inefficient, because they are costs that could and should have been avoided or minimized with more stringent regulatory and managerial oversight. Normatively, not only are these "mistakes" bad because they are inefficient, they are also bad because the nuclear regulatory system has institutionalized a mismatch between liability and responsibility. Generally, the persons or entities responsible for the mistakes are not liable. Moving away from the center of the discussion, the significance of the abandonment-costs problem for nuclear power and energy policy is discussed more generally. Further away from the center, the narrow but significant topic of abandonment costs is used to examine the consequences of a joint government-industry policy choice on law and legal institutions.

Energy law and energy policy are both dynamic topics. Energy policy is seen as fragmented, splintered, internally inconsistent, and indeterminate. Energy law is seen as ever-changing, elephantine, chaotic, and ineffective.[12] An objective observer easily can become a policy-making nihilist given these characterizations and walk away from a discussion about energy law and policy because it is overly complex and deeply uncertain. Such disengagement does not leave a very satisfying aftertaste. Energy industries account for a minimum of 10 percent of the GNP. Energy companies are among the largest entities in the private sector. Twelve of the first twenty companies on the Fortune 500 list are energy concerns.[13] The size and number of government regulators are comparable. The DOE, with 20,000 employees, and the NRC deal exclusively with energy matters. Other agencies spending considerable time on energy include the Department of Interior, the Department of Transportation, the Interstate Commerce Commission, and the Environmental Protection Agency, to name some of the largest. In addition to these agencies, committees and subcommittees of the House and the Senate and the caseload of the courts are also heavily energy-involved. Energy is not a self-regulating field; it taxes the time and resources of each branch and level of government. Energy law and policy are international in scope, historically rooted in our culture, and geopolitically entrenched in our governmental system. Energy is a matter of major concern in both the private and public sectors and at state and national levels. Nuclear power and abandonment costs touch all of these issues.

1. Institutional Setting

REGULATORY HISTORY

The transformation of commercial nuclear power and its regulation is the product of its peculiar history. The market for nuclear power, the regulatory institutions designed to promote it, and the current abandonment-costs predicament are consequences of a government-industry joint venture that failed. That failure triggered the transition. The earlier pronuclear policy was aided and abetted by a complex legal apparatus, which can best be explained by briefly reviewing the statutory scheme and judicial cases that implemented it.

The federal government has been pivotal in the development, regulation, and promotion of nuclear technology since its inception. Although subatomic physics has been on the cutting edge of the hard sciences since the turn of the century, the translation from a theoretical and experimental science to an applied technology did not occur until the United States government galvanized a preeminent group of physicists behind the design and construction of the atomic bomb. Under General Leslie Groves, the Manhattan Project was the coordinated effort bringing theory into actuality and culminating in the successful bomb explosion at Los Alamos.

Nuclear fission, the splitting of the nucleus of an atom with a consequent release of energy, is used in all commercial nuclear power reactors today. The first nuclear reactor in the United States was operated in 1942 by a group of scientists led by Enrico Fermi and Leo Szilard. The fission reactor was created by the United States to counter the perceived threat of Germany's building and using an atomic bomb. Though German scientists had discovered nuclear fission in 1938,[1] there was no real chance that Germany would use the bomb during the war. Nuclear power's first public appearance resulted in the desolation of two Japanese cities, the end of World War II, and the dawn of the Nuclear Age. The destructive force of the atom became known before the public was aware that the power could be tamed for such peaceful purposes as the generation of electricity. The end of the war ended the military's near-exclusive control of nuclear technology.

The early shift from military to commercial use was made at the behest of physicists intrigued with the scientific and technological mysteries of nuclear power. For many scientists, work on the Manhattan Project was the highlight of their careers.[2] For others, the movement from war to peace was a way of atonement.[3]

Although military control was looked upon with suspicion, the federal government was not removed from the regulatory process. To the contrary, the federal government steered the course of this technology through its infancy. The Atomic Energy Act of 1946 formally shifted control over nuclear development from the military to the civilian government.[4] The 1946 act attempted to keep secret all information about the development of nuclear power so that other countries would not be able to build a nuclear bomb. The attempt at secrecy failed when both the Soviet Union and Great Britain detonated their own nuclear devices. In addition, the act strictly maintained the government monopoly over the control, use, and ownership of nuclear energy until the Atomic Energy Act of 1954.[5]

Two regulatory bodies were created by the 1946 act. The civilian five-member Atomic Energy Commission (AEC) was the primary administrative agency. The chief functions of the AEC were to encourage research and promote development of the technology for peaceful use. The act also created an eighteen-member Congressional Joint Committee on Atomic Energy (JCAE). This watchdog committee comprised members from each legislative chamber. After the removal of the military from the process, the only persons with nuclear sophistication were the scientists who worked on the bomb, and AEC policy reflected their interests together with those of the JCAE.[6]

Very little development of commercial nuclear power occurred during the period 1946-1954. The physicists, naturally, were more involved with scientific problem solving than with commercialization. In the late 1940s and early 1950s, the AEC together with the JCAE shifted nuclear policy to producing electricity for the public's use on a larger scale. A small "breeder" reactor first produced electricity in 1951, but the major breakthrough came when the Navy's submarine Therman Reactor I began producing electricity in 1953. Under Admiral Rickover's direction, the groundwork was laid for the prototype of the present-day reactor, designed as part of the U.S. Navy's submarine program.

While the AEC and JCAE looked to the eventual commercialization of nuclear power, the Atomic Energy Act of 1946 restricted ownership of reactors and fuels to the government. By 1953, the Eisenhower administration, under pressure from scientists, business leaders, and diplomats, revised the nation's atomic energy policy and encouraged private commercial development through passage of the Atomic Energy Act of 1954. The 1954 act ended the federal government's monopoly over nonmilitary uses of atomic energy. It allowed for private ownership of reactors under an AEC licensing procedure. The policy of the new law was stated in a House of Representatives report, which stressed:[7]

> The goal of atomic power at competitive prices will be reached more quickly if private enterprise, using public funds, is now encouraged to play a far larger role in the development of atomic power than is permitted under existing legislation. In particular, we do not believe that any developmental program carried out solely under governmental auspices, no matter how efficient it may be, can substitute for the cost-cutting and other incentives of free and competitive enterprise.

The 1954 act, the bulk of which governs today, set the tone and the goals for commercial nuclear energy. Private-sector public utilities were designated to take the lead and to run the reactors. At the time, utilities believed that nuclear-generated electricity would be "too cheap to meter," that costs would be so low that they would need not bother billing customers. The peaceful use of such destructive resources would help absolve the guilt of Hiroshima and Nagasaki while keeping the United States in the forefront of the development and control of nuclear technology. This approach consolidated public opinion behind nuclear power.

Lewis Strauss, chairman of the AEC, interpreted the policy behind the 1954 act as a mandate to rely principally on private industry to develop civilian reactor technology. The first step, the Power Reactor Demonstration Program of 1955, was an attempt to involve private industry in a competitive program whereby five separate reactor technologies would be tested. Government and private industry were to develop reactors jointly. Once the reactors were developed, government was to step out of the project, and privately owned utilities were to assume fiscal responsibility. Private industry was not receptive to bearing the financial burden and was unenthusiastic.

The results of the Power Reactor Demonstration Program in its initial years were not overwhelming. Private firms were unwilling to invest in nuclear plants without government's shouldering financial responsibility. There was great pressure on the government to finance the industry, and Strauss felt that financial-liability roadblocks should be removed by government. The critical impediment was the nuclear accident. Officials of General Electric, one of the major reactor builders, threatened withdrawal from nuclear development activity, stating that GE would not proceed "with a cloud of bankruptcy hanging over its head."[8] In reaction, Congress passed the Price-Anderson Act of 1957, limiting industry liability and assuring some compensation for the public.[9] The act removed the last obstacle to private participation. Westinghouse executive Charles Weaver recalls, "We knew at the time that all questions (about safety risks) weren't answered. That's why we fully supported the Price-Anderson liability legislation. When I testified before Congress I made it perfectly clear that we could not proceed as a private company without that kind of government backing."[10]

Congressional hearings on the Price-Anderson Act reveal that there would be no commercial nuclear power plants built by the private sector without a financial safety net provided by the government. The act limits a public utility's financial exposure in the event of a nuclear incident. The ceiling for liability was set at $560 million in the original act. This amount consists of all the private insurance the utilities could raise, which from 1957 to 1967 amounted to $60 million, with the government standing good for the remainder. Every ten years the act comes up for renewal. Now, the act requires the utilities to foot the insurance bill. Under the 1975 amendments of the act, industry is assessed $5 million per reactor. There are eighty reactors, which together with $160 million of available private insurance equals a $560 million contribution by industry, essentially eliminating government participation.[11]

The Price-Anderson Act's $560 million limitation on liability is a hard, maybe even a "tragic," policy choice.[12] As noted earlier, government estimates of damages caused by a core meltdown are $14 billion. Already, TMI's costs exceed $1 billion. A $560 million liability limitation means that once that amount is reached, additional costs incurred as a result of a nuclear incident will be absorbed by the victims. The people who live near the plant may suffer personal and property damage in excess of the ceiling amount absent either a voluntary contribution by industry or an additional commitment by government. The government subsidy enables utilities to build plants without the normal checks against putting a defective product on the market. Such insulation from liability seems unfair, and may well be, but is entirely legal.[13] The Price-Anderson Act typifies the nature of nuclear power regulation. Government and industry have encouraged each other to participate in a long-term joint venture without assuming normal market risks. Instead, most risks are imposed on the public.

The first nuclear reactor to be connected to an electric distribution system in the United States began operating in late 1957 at Shippingport, Pennsylvania. Its sixty-megawatt capacity was the largest at that time. Over the next three to four years, larger and larger plants were built as part of the nuclear power experiment. The public and the electric utilities were becoming comfortable with the nuclear idea.

Electric utilities did not start ordering reactors in any number until manufacturers guaranteed plant prices. Reactor vendors, the manufacturers, needed to induce utilities to buy their product. The inducement came in the form of "turnkey" contracts, which placed all cost risks on the vendors. As an example, General Electric entered fixed-price contracts for the design, manufacture, and construction of nuclear power plants. Once a plant was built, regardless of cost overruns, the keys to the plant would be turned over to the utility. The first of these "turnkey" plants was a 650-megawatt plant built by General Electric for Jersey Central Power and Light. It was ordered in December 1963. There were nine turnkey plants, including Jersey Central's, with total cost overruns of between $800 million and $1 billion.[14] While the manufacturers incurred these large losses, the projects served their intended purpose of creating a market for nuclear power plants.

The relationship between manufacturer and utilities under turnkey contracts symbolizes the entire nuclear framework. An abnormal market was created when utilities willfully relied on manufacturers to guarantee the costs of their plants. The manufacturers viewed the turnkey contracts as loss leaders, later opening the nuclear market door to the emerging industry. Limited market risks and lessened fiscal responsibility were accepted by the parties to turnkey contracts because the government stood ready and able to absorb costs and protect the parties from extraordinary losses. When manufacturers assumed all risks until plants came on line, utilities were isolated from responsibility for construction costs in the same way the Price-Anderson Act insulated the industry from liability risks. The practice of insulation is also contained in the ratemaking process, in which ratepayers sustain utilities' decisions to overinvest in nuclear plants. This pattern of feigned dependence, risk insulation, and readiness to shuffle losses on to other parties continues today and contributes to the current abandonment-costs crisis.

The confidence created by ordering turnkey plants encouraged utilities to forge ahead into the nuclear power plant market. To promote this new market, manufacturers cited their ability to build plants and persuaded utilities that nuclear fuel was cheaper than fossil fuels. Although untested, with these assurances, numerous electric utilities began entering into "cost-plus" contracts for nuclear power plants. In contrast to the turnkey contract, the utilities agreed to pay for cost overruns. In what has been called the "Great Bandwagon Market" between 1965 and 1968, forty-nine plants totaling almost 40,000 megawatts of capacity were ordered under cost-plus contracts.[15] By the beginning of 1970, none of the plants ordered after 1964 were in operation. The total generating capacity of those in operation was only 4,200 megawatts, a little over four times greater than the average size of one reactor of those on order, compared with the 72,000 megawatts on order or under construction. In fact, the first reactor built without any government assistance did not begin operation until 1967. The industry's early years are based on wishful thinking that electricity would stay cheap, blind faith that the technology could be carefully watched, and unquestioned reliance on the hope that growth in the demand for electricity would continue.

What caused the surge of faith in nuclear power? First, the large growth rate in demand for electricity, which averaged 7–8 percent per year from 1960 to 1972, was not expected to subside.[16] Second, it was possible to realize economies of scale by building large plants (over 1,000 megawatts) and spreading the capital costs over a large amount of generated electricity.[17] Third, the technical results of earlier smaller reactors were encouraging.[18] A fourth reason, given by Bupp and Derain, was the lack of a credible challenge to nuclear power. The technicians involved in the creation and the development of nuclear power were the only people knowledgeable enough to challenge it, and they were not in a position to do so.[19] After the "Great Bandwagon Market" there was a short lull in orders for reactors, but by 1970 the utilities were back in the market with full force. Between 1970 and 1974, 145 reactors were ordered.[20]

Given a climate of limited responsibility, health, safety, and environmental concerns were partially anathema to the regulators. A casual attitude toward environmental health and safety remained until Congress passed the National Environmental Policy Act (NEPA) requiring environmental impact statements for all major federal activities.[21] Even then, the AEC's response to NEPA was to resist its force unless ordered to comply by a federal court or Congress. A federal court eventually held that NEPA's provisions applied to the AEC,[22] which subsequently drafted its own environmental provisions.[23] Subtly, the temper of regulation changed. People were no longer complacent with nuclear power safety and environmental claims made by industry and government. Even though Congress left the 1954 enabling legislation intact except for some 1964 amendments,[24] the consciousness raising accomplished through the environmental movement made people aware of radiological hazards.

Increasing public awareness of the dangers of radioactivity, however, was a frustrating strategy for nuclear foes. The demand for electricity was growing at an inviting rate; coal-burning facilities were an environmentally unattractive alternative; and with multiple oil-price hikes in the mid-1970s, nuclear power remained economically desirable. Additionally, because the preemption doctrine centralized decision-making power over radiological matters in Washington, government and industry could easily continue their pronuclear policy, ignoring safety and environmental questions. Prior to the mid-1970s, this promotional posture was solidified as the financial community further downplayed antinuclear forces by happily investing capital in what was considered a safe bet. Yet, in the mid-1970s a rift in the government-industry relationship started to develop, reflecting the trade-off between safety and finances and the conflict between politics and markets. The rift was first manifest in a bureaucratic realignment.

The AEC had conflicting functions, promoting the use of nuclear technology and insuring that the technology was applied safely. The commission was to give "go" and "caution" signals simultaneously. In 1974, realizing the cross-purposes of both promotion and safety oversight, Congress split the agency, creating the Nuclear Regulatory Commission (NRC), responsible for safety and licensing, and the Energy Research and Development Administration (ERDA), responsible for the promotion and development of nuclear power.[25] ERDA was later absorbed by the Department of Energy. The division and dissolution of the AEC had little immediate impact on nuclear power, because the promotional pattern had been set, and the NRC continued its predecessor's task. Furthermore, the split did not eliminate a basic regulatory conflict. The NRC has a promotional responsibility through its licensing process and, at the same time, must oversee plant safety. If the commission too vigorously exercises its safety role, then the attendant compliance costs act as a disincentive to invest in nuclear plants and cut across the NRC's promotional grain. The promotional emphasis continued vigorously until TMI. After 1979, the NRC increased safety inspections, stepped up enforcement, developed back fitting and emergency-preparedness rules, and adopted a more safety-conscious attitude. This attitude was more costly to the industry and was a clear, but temporary, departure from past practices.

This brief institutional history highlights the fundamental elements of nuclear power regulation. The government, in its haste to move nuclear technology into more peaceful commercial uses, forged a strong and protective partnership with key private-sector actors: equipment vendors, plant contractors, and utility owners. Thus, the principal private- and public-sector participants were interested in promotion and were concerned little with harmful externalities. Industry's interest in profit maximization and government's interest in expanding commercialization thwarted a cautious approach to environmental and safety problems associated with nuclear electricity generation. The financial arrangement between government and industry reinforced a promotional policy. The private sector was insulated from liability on normal market risks, and the government was not financially accountable insofar as any losses ultimately attributable to them could be spread to taxpayers. This state of affairs changed after TMI inaugurated a regulatory transition preoccupied with the financial health of nuclear utilities.

NUCLEAR POWER AND THE COURT

This section briefly describes a change in judicial attitude toward nuclear power regulation by concentrating on a few United States Supreme Court decisions. These opinions do not canvass the whole area of judicial review of nuclear power issues. However, they clearly reveal a change in judicial temperament.

The few early cases on nuclear power regulation reflect a lack of concern about energy, environmental, and safety costs and depict a faith in the central administration of the industry. The Supreme Court first spoke about nuclear power in 1961 in a case contesting the Atomic Energy Commission's grant of a construction license for a breeder reactor plant.[26] The precise legal question was whether the Atomic Energy Commission must make the same definitive finding of safety before granting a construction license, the first in a two-step procedure, that it must make before granting an operating license in the second step. The commission issued the construction license with a safety analysis less detailed than required by the commission's own regulations and by Congress's statutory standard for the operation license. The federal court of appeals for the District of Columbia[27] held that both safety analyses had to be of comparable quality. The United States Supreme Court, in an opinion by Justice Brennan, reversed the lower court in favor of the AEC:[28]

> The Commission, furthermore, had good reason to make this distinction. For nuclear reactors are fast-developing and fast-changing. What is up to date now may not, probably will not, be as acceptable tomorrow. Problems which seem insuperable now may be solved tomorrow, perhaps in the very process of construction itself. We see no reason why we should not accord to the Commission's interpretation of its own regulation and governing statute that respect which is customarily given to a practical administrative construction of a disputed provision.

Justice Brennan also addressed the "fears of nuclear disaster which respondents so urgently place before us," also by deferring to the commission: "We cannot assume that the Commission will exceed its powers, or that these many safe-guards to protect the public interest will not be fully effective."[29]

The *Power Reactor* case is part of a consistent pattern in the fabric of administrative law. Courts defer to agencies for numerous reasons. The doctrine of separation of powers[30] and the system of checks and balances established by the Constitution support a policy of deference. Congress, a more democratic branch than the judiciary, created the AEC and delegated to it the authority and responsibility for administering commercial nuclear power. Thus, as a matter of constitutional law, courts grant agencies much leeway. Also, the agency is supposedly the specialist and has the expertise, hence the competence, to decide issues peculiar to its charge. The court simply is not well equipped to second-guess the expert body on factual or policy questions.[31] As long as the agency's action is not arbitrary, capricious, or unreasonable, and an agency's factual findings are supported by substantial evidence, courts are obligated by law to defer.[32] Consequently, the

Supreme Court's ruling reinforced the power of the AEC to direct the country's nuclear program.

The next major nuclear case deals directly with the question of preemption. In a federal system there are one central sovereign (the federal government) and fifty other coequal sovereigns (the states). Power sharing and cooperation among these sovereigns are central to the ideology of federalism. Nevertheless, some device is needed to break occasional regulatory deadlocks. The preemption doctrine roughly accomplishes this function by establishing a hierarchy of situations in which federal prerogative controls. The preemption doctrine emanates from the Supremacy Clause of the Constitution.

> This Constitution and the Laws of the United States which shall be made in Pursuance thereof; and all Treaties made, or which shall be made, under the authority of the United States, shall be the supreme Law of the Land; and the Judges in every State shall be bound thereby, any Thing in the Constitution or Law of any State to the Contrary notwithstanding.

Preemption can be triggered in two general ways. Congress can expressly or implicitly intend to preempt state law. Or, the scheme of federal regulations can be so pervasive that state intrusion into a federally regulated area conflicts with or impedes the functioning and purpose of federal legislative efforts.

The constitutionality of a Minnesota statute was at issue in *Northern States Power Co.* v. *Minnesota*.[33] The statute imposed more stringent requirements for the release of radioactive waste from a nuclear power plant than AEC regulations. In broad language, the federal appeals court held that:[34]

> [T]he federal government has exclusive authority under the doctrine of preemption to regulate the construction and operation of nuclear power plants, which necessarily includes regulation of the levels of radioactive effluents discharged from the plant.

The United States Supreme Court affirmed *Northern States Power* without opinion. Technically, the only thing that can be inferred from the affirmance is that at least a plurality of justices agreed with the lower court's result.[35] Thus, the Supreme Court set the practice of deference to a centralized nuclear agency—that is, until *Pacific Gas & Electric*.[36]

In 1976, the California legislature amended the state's Warren-Alquist Act and conditioned the construction of nuclear plants on the finding that adequate storage and disposal facilities are available for nuclear waste. The waste disposal problems of nuclear reactors are at crisis proportions. First, nuclear wastes are highly toxic. Harmful radioactive waste lasts thousands to hundreds of thousands of years.[37] Second, there are only two off-site waste storage facilities in the country. Third, the bulk of the waste is stored on the plant site. These sites are reaching capacity, and on-site storage is being expanded.[38] Finally, and most curiously, the federal government had no mechanism for resolving the waste disposal problem until 1982, when the Nuclear Waste Policy Act was passed.[39] By the terms of the act, five disposal sites are to be nominated no later than January 1, 1985,[40] and by July 1, 1989 three are to be recommended, from which the president is to choose one for a repository.[41]

California's legislative response to nuclear waste was to impose a moratorium on new nuclear plants until a disposal method is found. The Pacific Gas and Electric Company, a utility that builds nuclear plants, brought suit alleging that sections of the state legislation were preempted by the Atomic Energy Act because the state-imposed moratorium completely frustrated the federal effort to develop and promote the use of nuclear power. The district court agreed with the utility. The court of appeals reversed, reasoning that the Atomic Energy Act constitutes a congressional authorization for states to regulate nuclear power plants "for purposes other than protection against radiation hazards."[42] The court held that California could protect its financial flanks as the "other purpose," because "uncertainties in the nuclear fuel cycle make nuclear power an uneconomical and uncertain source of energy."[43]

Justice White's opinion, upholding the appeals court's denial of preemption and sustaining the constitutionality of the California statute, is based on a distinction between radiological hazards, which fall in the province of the federal government, and nonradiological matters, which are to be controlled by the states. The Court listed traditional areas of state responsibility as including the need for power, electric reliability, economic feasibility, land use, and ratemaking.

A moment's reflection on the distinction relied on by the Court should reveal the significance of the opinion. The distinction is, in large measure, spurious. Safety and finances are not discrete topics. Waste disposal is a radiological hazard as much as it is an accounting entry on the utility's books. The opinion is less a departure from the letter of the law than it is a departure from a well-entrenched mindset[44] favoring a strong, central, federal regulatory regime. The Court's sleight of hand was the characterization of California's nuclear moratorium as a financial rather than a safety consideration. Fact manipulation and rhetorical characterization are tools of a lawyer's trade. The manipulation was made easy by the Court's unwillingness to delve into the motive of California legislators. As long as California maintained that the legislation was aimed at economic problems, the Court's inquiry ended absent a patent falsification.

The United States Supreme Court, for the first time, ruled explicitly that decisions regarding nuclear power are not within the exclusive province of the federal government. Significant decisions about this most controversial of natural resources are also to be made by individual states. One commentator has written that such a result is "unsurprising"[45] because of the express language in the Atomic Energy Act of 1954 reserving state decision-making power "for purposes other than protection against radiation hazards."[46] Most commentators,[47] as well as the bulk of court decisions,[48] however, relegated virtually complete decision-making power to the federal government and conceded that the field had been preempted. Indeed, the Court easily could have justified an opinion holding that the California statute was preempted either by stating that safety and financial matters are indistinguishable or by noting how pervasive federal regulations are in the nuclear field and how the amendments to the Warren-Alquist Act frustrate federal controls. *Pacific Gas & Electric* signifies an important departure from previous nuclear regulation and announces a period of shared decision making between federal and state governments, a period of cooperative federalism.

The Supreme Court is clear in its line of demarcation as an abstract legal rule. Matters radioactive are delegated to the federal government, and "other" matters are within the control of the states. Even more narrowly, finances are seen to be of local rather than federal concern. State legislators effectively can ban the construction of nuclear power plants if, in their opinion, the nuclear alternative is unsound financially. As a matter of energy policy, things are not quite so simple. The direct effect of sharing decision-making power is to fragment a national nuclear policy, and with it a comprehensive national energy plan. The practical consequences are even less clearly delineated. How, for example, does a decision maker differentiate between radioactive and financial issues? Is waste disposal, transportation of nuclear materials, or plant decommissioning radioactive or financial? What is the real motive behind a legislative or bureaucratic decision to slow down or halt nuclear power development in a state? The relegation of decision-making authority to the states means that they can (and from the perspective of the citizens of any given state must) make decisions along parochial lines. States fearful of the specter of nuclear power can restrict its use or impose a construction ban. Less populous states may foster the nuclear option. *Pacific Gas & Electric* is premised on a political choice favoring state participation in decision making. Shared decision making has fragmented policy, has diffused power, and has dislocated the national market by raising costs unevenly.

In one sense, the Supreme Court's pronouncement merely adds a gloss on the previous structure of decision making. State public utility commissions and public service commissions already have the responsibility for the financial destiny of nuclear energy, because they are the principal decision-making bodies regarding rate levels utilities can charge for their nuclear-generated electricity. Simplistically, if rates are not high enough, the utility has a disincentive to invest in a nuclear plant. The Court's imprimatur clears the way for state legislation on "other" matters free from fear of preemption and introduces an important political element into nuclear policy formation.

If TMI serves as the benchmark for the change in economic and political climates, then *Pacific Gas & Electric* serves as the benchmark for legal change. Together, TMI and *Pacific Gas & Electric* signify two essential characteristics of the transition. The first characteristic, exemplified by TMI, is the emphasis on financial and market needs rather than on safety matters. Although public attention is drawn to the financial fallout of contemporary nuclear policy, safety issues are not solved or of secondary concern. Safety is integrally connected to financial matters. Health and safety issues must be acknowledged in assessing ways to ease the financial impact of abandonment costs. Rising expenses due to plant decommissioning or waste disposal, as examples, are directly attributable to health and safety variables. Similarly, efforts to streamline the regulatory process or to facilitate plant construction should not be established if health and safety risks are increased. Therefore, to focus on the financial weakness of the industry is not, and should not be, a substitute for health and safety regulations.[49] Such a trade-off is a macabre pas de deux, made only more frightening by the interconnection between nuclear power and nuclear weapons.[50]

As noted, the TMI accident was close to a core meltdown; therefore, the change in emphasis from safety to finances is paradoxical. However, the paradox can be explained. Moving the debate away from polemical, frightening claims of safety risks, risks for which virtually no hard data exist, toward a debate about dollars and about where dollar costs should be spread is oddly assuring. The financial dislocation caused in the industry underscores the need for a more careful approach to promoting nuclear technology. More important, safety and finances have a curious relationship. Politics are introduced at the intersection of safety and finances. No system is fail-safe. Determining safety levels and allocating costs through a regulatory system are essentially political. Health, safety, and financial issues are best resolved through a politically responsive decision-making process. Financial and regulatory communities are beginning to recognize a new political configuration as they engage in the process of spreading abandonment costs with an eye on the safety aspects of nuclear power. Currently, decision-making power is diffused in response to political pressures. Nuclear decisions are being made not only in Washington but in the states, as well. The new era of nuclear power regulation is thus distinguished by another, equally important characteristic—the decentralization of decision-making authority.

Decentralization is clearly exemplified by *Pacific Gas & Electric*, which explicitly recognizes the states' role in nuclear decision making. The move toward decentralization had to overcome an important history of centralized decision-making by the federal government. Public disillusionment with nuclear power was born from a sense of frustration. Determining safety risks and environmental and health consequences of this energy source is difficult. Hard, quantifiable data are often absent, speculation is rampant, and most of the information is controlled by the industry. Decision-making power had been centralized in Washington, D.C., and industry enjoyed a cozy relationship with regulators.[51] Citizens had little or no effective voice in the central regulation of this complex industry. One source of hope for a voice in nuclear policy making lay with state government. Theoretically, state legislatures should be more responsive when it is too costly, cumbersome, and difficult for individuals to organize and lobby Congress. However, there exists a significant impediment to a successful state response to nuclear power—the United States Constitution. Even when antinuclear forces have been successful in converting state legislators to their cause and in passing industry oversight legislation, constitutional law is such that federal regulations take precedence over state laws. Federal preeminence had been especially true in the nuclear field.[52]

Both the early history of nuclear regulation, described in the last section, and the nature of the nuclear industry make a persuasive case for centralized regulation of commercial nuclear power. This complex, costly, dangerous technology could not be left to a military establishment primarily interested in weaponry. Nor could the free market be relied upon to transfer the militaristic technology to commercial use. Nuclear power was too untested and too unknown to rely on private actors to internalize safety and environmental externalities. Moreover, the private sector did not want the financial risks. Government support and oversight were necessary. The industry, a large, complex, controversial, and publicly sponsored enterprise, could exist only with the assistance of a concentrated, bureaucratic decision-making body. Consequently, legislation creating the commercial nuclear regulatory bureaucracy centralized decision making in Washington for the smooth shift of nuclear technology from military to commercial use. This body, formerly the Atomic Energy Commission, now the Nuclear Regulatory Commission, is an administrative agency of the federal government and is in a position superior to that of the states to promote the uniform use of nuclear energy. Centralized federal decision making made the development of commercial use possible. Centralization has also contributed to the downfall of sound nuclear policy in the 1980s.

For an industry as complex and controversial as the nuclear power industry, a pronuclear consensus existed until the mid-1970s. Energy costs were cheap until the Arab oil embargo of 1973, and there was little public consciousness of environmental matters until the late 1960s. Also, with public attention in the early and middle 1960s turned to civil rights, confidence in government did not wane until the height of the antiwar movement in the late sixties and the Watergate debacle.[53] Doubts about the wisdom of investment in nuclear power and doubts about the federal government's ability to monitor the nuclear industry were not part of the country's collective psyche.[54]

Thus, *Pacific Gas & Electric* constitutes the second characteristic of the transitional period of nuclear power regulation. What was once almost exclusive control of the industry by the federal government is now shared by the states.[55] *Pacific Gas & Electric* is not a random case or a fluke. The language affirming state power is clear and crisp. The bow to cooperative federalism, and therefore decentralization, was repeated in *Silkwood* v. *Kerr-McGee Corp.*[56] where the Court held that a state's punitive damage award for radiation exposure was not preempted by federal law.

An intractable given of our legal system, and more generally of our language and culture, is that rules are malleable. The Court has greatly altered the face of nuclear regulation and the shape of the nuclear industry without overturning either legislation or precedent. The Court's opinion embodies the contemporary perception of nuclear power. TMI's financial fallout and *Pacific Gas & Electric's* cooperative federalism have forced industry and government to rethink the role of nuclear power in society, to reevaluate the nature of their collaborative relationship, and to reorient the regulatory climate.

The modern, or, perhaps in keeping with contemporary scholarship, postmodern, era of nuclear power regulation is therefore characterized by two phenomena. First, the focus is on the financial heart of the industry rather than on the complex and controversial safety aspects. Second, decision-making power is decentralized by de jure as well as de facto recognition of the states' role in nuclear energy law and policy. As a matter of institutional structure and intentional design, nuclear decision-making authority is diffused. These two phenomena are indicia of a changing, more cautious regulatory attitude toward the industry. A subtle normative conflict exists between the two indicia. Finances are generally arranged by markets guided by an efficiency criterion. Power is generally allocated by politics motivated by a concept of fairness. The regulatory state mixes both sets of variables. In contemporary nuclear regulation, efficiency and fairness conflict, as will be demonstrated in chapter six. During the transitional period, policy disarray occurs because an incongruity exists between the needs and desires of political and market actors. Now the legal system must be retooled to accommodate these conflicting interests.

NUCLEAR POLICY

Although nuclear power plays a central part in the country's energy planning, economic and financial downturns have halted its growth and development temporarily. Just as the nuclear industry is experiencing a readjustment in the private-market sector, the industry must also experience a regulatory political readjustment. Nuclear power must be placed into a larger economic-energy context for us to appreciate its position in our energy future and to begin to understand the needed regulatory response to the current instability in the industry.

Every presidential administration since Franklin Delano Roosevelt's has had a significant role in shaping nuclear policy. Since Roosevelt's decision to build the bomb and Truman's decision to drop it, since Eisenhower's maneuver to shift nuclear technology from military to civilian use and Kennedy's decision to open up private control over uranium, nuclear policy has been greatly directed by executive action.

The "energy crisis" hit the 1970s with great force and found the country unprepared. President Nixon's Project Independence[57] set the tone for the decade by declaring energy security a domestic priority, but it did little more than initiate the formulation of a national energy policy.

No president has had a more dramatic impact on energy policy as a whole than President Carter. President Carter built his administration, then lost it, on the energy crisis.[58] His administration, in response to the economic and national security threats engendered by the Arab oil embargo, institutionalized energy regulation in ways no other administration ever contemplated. Carter's most notable legacy was the attempt, and partial success, at coordinating federal and state governments behind a comprehensive national energy program. The Carter administration saw the creation of a major energy bureaucracy, the Department of Energy, and the redefinition of several major energy-related agencies.[59] More important, that administration also was responsible for steering massive legislation through Congress specifically designed to project a national energy plan.[60]

Carter's energy addresses were instrumental in bringing to public awareness the magnitude of the "energy crisis" and its many delicate facets. No longer is the energy crisis in the public consciousness. No longer do phrases such as "moral equivalent of war" capture our attention. Yet, the problems, predominately finances and national security, giving rise to the atmosphere of crisis have not disappeared. The massive legislation and bureaucracy established by the Carter administration have not evaporated, despite President Reagan's explicit program to abolish energy agencies and his informal undermining of the energy bureaucracy by appointing people committed to dismantling the regulatory structure. Nor have future prospects of further "crisis" dissipated. A fragmented national energy plan is dangerous for national and world economies. Falling oil prices in the mid-1980s threaten to recreate the energy crisis of the early 1970s, thus reminding us of the need for or desirability of a national response.

Aside from the heightened rhetoric equating the energy crisis with a spiritual crisis, Carter's April 18, 1977 energy address contained an important truth about energy policy. The theme of the speech was that our country, indeed the world, was experiencing a transition in energy forms. The president spoke about three transitions. First, the transition from wood to coal presaged the Industrial Revolution. Second, the growing use and dependence on oil and natural gas revolutionized global transportation. This last period has begun to end. The era of cheap energy is over. The end is in sight for nonrenewable resources, and new "alternative" forms of energy must be sought. The third transition, then, is from nonrenewable to renewable resources. Although Carter did not make an explicit connection between this third energy transition and nuclear technology, nuclear power may be the core connection of the third transition. This transition is only beginning; the parameters of the period are taking shape, and the larger contours are perceptible. During this period new sources of energy must be sought, new technologies assimilated, and new decision-making systems explored. Our society finds itself in the peculiar and uncomfortable position of describing a future energy policy while redefining the relationship between government and industry.

That nuclear power is part of this transition is no historical accident. The current energy climate is dependent on two basic energy forms, petroleum and electricity. There is limited cross-substitutability between these two forms. Although oil and natural gas can be used to burn boilers, creating steam to turn turbines to generate electricity, the country does not now devote precious oil and gas reserves for that purpose. Oil is better used in the transportation sector. Electricity, because it cannot be stored, has little usefulness for transportation outside inner-city mass transit and is therefore used mainly for heating, cooling, lighting, and industrial purposes. Added to these realities, oil and gas reserves are getting harder and harder to locate, explore, and develop. Pursuant to the laws of the invisible hand, increased scarcity means that oil and gas are increasingly more expensive. In 1970, petroleum sold for $6.70 per barrel. It climbed to a high of $39.30 in 1980, a fivefold increase within a decade; sold for about $29 per barrel in 1984[61]; and sells for a low of approximately $10 in 1986. Natural-gas prices have skyrocketed similarly. During the same period a thousand cubic feet of natural gas went from $.53 to 2.73 mcf.[62] These resource cost statistics signal the attraction of nuclear power. If nuclear is cost-competitive and nearly inexhaustible, then it can assume a transitional role as the country moves to renewable energy even as oil and gas prices stabilize with a perceived dislocation in OPEC. This portrait is not without blemishes.

Recent estimates of capital investment for electric generation show nuclear power to be more costly than coal, oil, or gas. The cost of constructing a nuclear plant is $3,900/kwh, compared with $2,700-2,900/kwh for coal, $2,300/kwh for oil, and $1,700/kwh for natural gas.[63] The most likely substitute for nuclear-generated power is coal, because of its abundance. Nevertheless, coal is a very problematic resource. The risks of nuclear power lie in contemplation of a low-probability, high-risk event. Translated, such speculation means that we simply do not know the health and safety effects of a nuclear accident, because, fortunately, we do not have the necessary data and experience to test conflicting hypotheses. Greater coal use carries similar risks. Even though the number of miners with black lung disease and accidental deaths is knowable, uncertainty surrounds the horror of atmospheric cataclysm due to CO_2 build-up or of ecological destruction due to acid rain. Coal, therefore, is not an easy choice. Like nuclear, coal has promise because it is plentiful and, given certain assumptions, is cost-competitive. Unlike nuclear, coal is more readily exhaustible. It is nuclear power's abundance that makes it such an important transitional resource.

In Nixon's Project Independence speech, nuclear power was to provide 30-40 percent of our energy needs by the end of the 1980s and 50 percent by the year 2000.[64] President Ford continued Project Independence, calling for a ten-year program to build two hundred additional nuclear plants.[65] And, *Energy Future*, the widely read report by the Harvard Business School, attributed a 7 percent share of all energy production to nuclear power.[66] Currently, nuclear power generates approximately 13 percent of our electricity and about 4 percent of all of our energy needs. Nuclear power was intended to play such a prominent role because it was cost-competitive and because safety risks were discounted. The higher price of electricity generated from power plants using oil increased the financial attraction of nuclear power. However, the demand for electricity, which was increasing at 7–8 percent per year in the 1960s, fell to a level of 1–3 percent per year by the late 1970s and early 1980s, with a 2 percent decline in 1982. High electric bills and the recession caused many persons to conserve. Economic forecasters were surprised at the elasticity of demand for electricity relative to price hikes. The immediate pressure to increase capacity had vanished, and projections for more plants into the turn of the century had diminished. Demand was down, prices were up, and utilities were running well below capacity.[67] Demographic uncertainties such as population growth and migration patterns made demand projection difficult. Market uncertainties contributed to an unstable period for energy policy development.

The nuclear power industry began its present struggle in the 1970s. While the demand for electricity was going down, the costs of building a nuclear reactor were going up. The AEC was required to conduct detailed evaluations of the environmental impact of nuclear plants before issuing licenses. The evaluations added to delays in construction and higher costs. In addition, safety was also becoming a decisive issue after TMI. The NRC, pressured by public and special interest groups, began to reevaluate the nuclear power industry. Nuclear policy was now buffeted by political as well as market winds. As nuclear power entered the 1980s, both financial constraints and public concern about safety made it a less than desirable choice as a viable alternative to oil or coal. Because of the lower-than-expected increase in demand for electricity, it became increasingly difficult to rationalize nuclear power as economical. There was a "de facto" moratorium on nuclear power plant orders, which continues today. The moratorium was instigated by the market and reinforced by political pressures growing increasingly skeptical about the wisdom of nuclear power. Policy was changing as a result of political and market forces. Yet, large traces of the previous centralized promotional policy remain in the legal system assigned to make decisions about nuclear energy. Specifically, legal decision makers are allocating abandonment costs, and those decisions are informed by institutionalized policy preferences favoring the nuclear industry.

By the end of 1983, over one hundred nuclear power plants had been canceled. Electric utilities have been canceling not only plants where construction has not begun but also many plants in which billions of dollars have been invested, bringing the total bill for expenditures on discontinued nuclear plants to between $10 and $100 billion.[68] The following examples illustrate the magnitude and breadth of the crisis.

In the summer of 1983, the Washington Public Power Supply System (WPPSS), a consortium of public utilities in the Pacific Northwest that had planned to build five nuclear power plants for $4 billion (an estimate that would eventually rise to $24 billion), defaulted on municipal bonds used to finance the venture. WPPSS is the largest default in U.S. history and sent shock waves throughout the financial community. Only one of the five plants is to be completed. WPPSS has become a financial disaster, with bankruptcy threatening.[69]

In January of 1984, Commonwealth Edison Company, the Chicago electric power utility, was expected to receive a license to operate its Byron nuclear power plant. Commonwealth had one of the most successful nuclear programs in the country. It had seven plants in operation, one being started up and four others under construction. The company built its plants at a lower cost and quicker pace than other utilities and also employed a large group of experienced nuclear plant managers. In spite of its successful past history and experience, the NRC initially denied the operating licenses of two Byron reactors because Commonwealth had not assured the quality of construction at the plants. The denial surprised many in the nuclear power industry, because of Commonwealth's experience.

The Public Service Company of Indiana announced a few days later that it was canceling all further work on its two reactors at Marble Hill. The plant was expected to cost $4.5 billion to complete in addition to the $2.8 billion already spent. Recently, there has been talk of converting Marble Hill to a coal-fired power plant.

Cincinnati Gas and Electric and its two partners also announced in late January 1984 that they would be getting out of nuclear power. The partners' Zimmer nuclear plant is to be converted to a coal-fired plant, because the cost of finishing the 97 percent completed plant was expected to double the $1.7 billion already invested. Estimated costs of converting Zimmer from nuclear to coal are $1.7 billion.[70]

There have been problems at other plants, as well, including some in operation. There was a mishap at the Ginna nuclear power plant operating in Rochester, New York in January of 1982. The Diablo Canyon plant in California is located close to an earthquake fault, which has kept it under continued reevaluation and, although completed, not yet fully operational. The Shoreham plant on Long Island, New York has had difficulty with the development of the required emergency evacuation plans and also has defects in the plant's generators, both of which have contributed to its being ten years behind schedule and billions over budget. And, the Seabrook power plant in New Hampshire is experiencing a financial crisis in the face of assessments that the plant is not needed and that any increase in demand can be satisfied with Canadian electricity. Other signs indicate that the costs of the industry have been underestimated. Many feel that waste disposal and plant decommissioning alone hold financial costs that have not been fully computed into the costs of electricity. The nuclear power industry has been struggling in the last few years to maintain a place in the future electric power structure of the United States.[71]

While the industry displayed early successes, its current problems have dominated attention. One simple (and extreme) solution presents itself. The country's nuclear policy can be dictated by the market, and the industry can be allowed to die a financial death. This resolution denies the history of industry involvement with government and ignores the political significance for future energy regulation. If the current financial ills of nuclear power are not treated also as political ills, nuclear power will enjoy a resurrection when market indicators are more favorable. Consequently, we will learn little about either the nature of the patient or the pathology of the illness. Nuclear policy must be studied as a manifestation of both politics and markets.

Complete abandonment of nuclear power is not a realistic alternative. Both American government and American industry have invested heavily in the nuclear power option. One author estimates a recent annual investment at over $400 billion.[72] This figure does not include nonquantifiable investments such as individual career choice, corporate management design, bureaucratic initiatives, and emotional and intellectual commitment. Neither government nor industry will either politely or quickly walk away from these investments. Our energy needs are such that new energy mixes are being tried, new sources sought, and new strategies considered. Although possible, it is simply not true that a no-nuclear future is a real probability. Nor is it true that nuclear power can exist unregulated.

As the nation experiments with different energy policies, it must also experiment with different forms of energy regulation. A new regulatory climate is as much a part of the transition as resource mix. The new regulatory structure will be formed during the transitional period as regulators grapple with cost-allocation decisions. These decisions necessarily will be influenced by contemporary political choices, which, in turn, will be reflected in the legal decision-making apparatus, thus repeating the cycle by which law follows policy.

TRANSITION

For now, and for the foreseeable future, nuclear power is not a financially attractive investment. The transitional period was earlier depicted by two phenomena, the shift in concern from safety to finances and the decentralization of decision-making power and authority. These characteristics form the basis for the regulatory reforms needed to spread costs fairly and efficiently and that will develop a prospective nuclear policy. How the legal system allocates costs will be reflected in monthly utility bills and will have an impact upon the viability, structure, and design of the electric utility industry and, consequently, upon the country's energy program.

The United States is every bit as dependent on an electricity-based economy as it is on an oil economy. The assertion about the impact on energy policy is not intended to be a hyperbolic proposition in the sense that an uninformed or wrong solution will bring energy or economic chaos. Recall, nuclear power accounts for only about 13 percent of the electric supply. Given lower demand and excess capacity, the difference can be regained in the absence of nuclear power. Indeed, looking at the most extreme case, the country could scrap nuclear power entirely and replace it with coal-fired plants, alternative sources, conservation measures, or some combination of the three and still not rely on oil or natural gas to fuel electric generators. Financially, the costs are not completely debilitating, and many can be spread to numerous groups over a period of years. Nonetheless, the no-nuclear option is unrealistic given the nature of our capitalist democracy. As idealistic as No Nukes advocates may appear, and as popular antinuclear sentiment increases, these forces are not particularly well represented in decision-making circles. They are less cohesively organized, more poorly financed, and less well received by policy makers, and, consequently, they exert a less powerful policy-making voice. Too much time, money, and effort has been committed to nuclear technology over the last five decades by the public and the private sectors, domestically and internationally, to make backing out of nuclear power economically or politically feasible.

Even though our country could survive without nuclear energy, a pragmatic justification of nuclear power does not remove the need for a significant reorientation of nuclear power regulation, decision making, and policy making. Thus, the question posed is not whether the country can or should do without nuclear power—it most likely can. What remains is to design a mechanism to resolve the current costs problem, because the nuclear dilemma is emblematic of the type of polycentric policy problems confronting society. That mechanism will also determine the contours of future nuclear policy.

The interaction of nuclear law and policy is such that the institutional biases, decision-making principles, and policy-making goals factored into the resolution of the abandonment-costs problem remain behind after the current problem recedes, and shape the form of policy for decades absent another "crisis" event. The decision-making system and the regulatory structure developed in response to the costs problem will become institutionalized and entrenched in our legal, political, and economic cultures. The institutionalized policy choices and structural arrangements that have heretofore existed are a primary cause of the problem of nuclear abandonment costs. The financial failure of commercial nuclear power is no more the result of pure market failure than it is the consequence of regulatory failure. Instead, the industry's state of poor health is better diagnosed as a case of regulatory/market failure during the advanced stages of capitalism in an activist state. This bulky description means little more than that the failed nuclear power venture is the failure of a joint government-industry enterprise looking for a place to spread the costs of unanticipated and unwanted risks.[73] The previously developed system has a method of allocating abandonment costs. If the system goes according to the script, then ratepayers will pay the bulk of the costs; utilities will have little incentive for cost avoidance; state and federal regulators will remain impatient, if not hostile, to consumer interests; and nuclear power will proliferate irresponsibly, maybe dangerously.

The unified nuclear promotional policy that once existed has ruptured, indicating a need for a new policy-making and decision-making structure. A tear in the regulatory relationship is manifested by a divergence concerning policy between government and industry. The divergence between government and industry is evident in a Congress that asks industry to absorb waste disposal costs, in the NRC, which requires utilities to pick up decommissioning and additional safety costs, and in state utility commissions, protective of consumer interests by disallowing abandonment costs in the rate base. The break is also evident in judicial decisions imposing liability on industry and in state legislatures that respond with industry-checking legislation. Further, the crack in the united front of the atomic fraternity is demonstrated by intramural fighting within the industry. The infighting goes beyond normal competitive posturing. Utilities are bringing multi-million-dollar lawsuits against architects, reactor vendors, and engineering contractors for design, manufacturing, and construction mistakes. Shareholders are suing management for breach of fiduciary duty for choosing the nuclear path and for continuing down that path despite financial danger signals. Likewise, large and small customers are suing utilities in the hope of preventing costs from being passed through to them.

The legal and policy-making system stands at an important crossroads during this transitional phase. Much of the crisis atmosphere is over as a matter of public consciousness, if not of policy choice. The respite gives us the opportunity to evaluate the success or failure of particular substantive positions, the role of the central government in the design of a promotional nuclear policy, and the responsibility of industry for its mistakes. The transitional period is one of retrenchment and rethinking, as well as a period with its own peculiar problems and characteristics. It is also a period of maturation. Chief among the signs of the period is a search for nuclear power's own identity. This period is one of turbulence. The once-strong faith in the power of the atom to solve energy problems has been shaken badly. The conscious acceptance of this lost faith is endemic to a full realization of an irreducible and intractable fact about law and legal institutions; they reflect human frailty as much as they are responsive to human aspiration. TMI is perfect evidence that the nuclear regulatory system is dependent on human awareness, that nuclear technology is not self-sustaining, and that the imperfectability of human nature is contained in the institutions we create. According to the Heisenberg Uncertainty Principle, the human observer quantitatively and qualitatively offsets scientific measurements. The human actor affects nuclear regulation no less than it affects nuclear physics. Accepting the human dimension of nuclear power is the single most important realization that society must face during the transitional stage. Human and political sensitivity is a prelude to a fuller understanding of nuclear regulation and can lead to a better, more responsive policy.

There is a deep issue at work during the maturation process that underlies a profound, possibly disturbing, irony. TMI was a safety accident and a near-catastrophe. Yet, we occupy ourselves with analyzing the financial and market aspects almost as if the image of a core meltdown is too horrible to contemplate. As a defense mechanism, we simply disregard the safety aspects and concentrate on something more quantifiable—the dollar. Instead of focusing on the softer issues of human frailty and lack of political accountability as possible causes of potential human annihilation, we turn our attention to the hard question of finances. The *Pacific Gas & Electric* case reinforces the move from safety to finances by relying on an economic/safety distinction. However, the more important political message of *Pacific Gas & Electric,* the decentralization of decision-making power, must not be ignored.

A note of caution is offered: Ideas contain the seeds of their own demise. The focus on finances opens debate about the place of nuclear power in our society. The danger is that by overly concentrating on fiscal and market implications, a seductive path is presented, leading to lower safety standards as a means of insuring the financial integrity of the utilities, a mistake that ignores the main lesson of the transitional period. If nuclear technology is only as reliable as the actors regulating the industry, and if the momentum of nuclear technology has carried us further than the regulatory state can currently handle, then perhaps we should be cognizant of the dislocation and redesign a regulatory system that can keep pace with the technology. The human and political dimensions must be factored into decision-making and policy-making processes rather than dissociated from them.[74]

A process by which complex technological decisions can be made must be established. The process approach for decision making and policy making in the nuclear area is warranted on several counts. First, the very nature of the problems and issues surrounding nuclear power are multivaried. In recent legal literature they are known as polycentric problems, because they are complex, contain numerous uncertainties, affect different interest groups to different degrees, and, in the case of nuclear power, are transgenerational.[75] As described, polycentric problems are as normative and political as they are scientific and technical. Second, and more significant, there is no fixed answer to many questions about nuclear power. Safety and risk assumptions are matters of degree. Nuclear plants will never be 100 percent safe—nothing is. Safety, together with cost issues, can be satisfactorily resolved only through the design of a politically and socially acceptable decision-making system.[76] Though the system necessarily will be imperfect in a scientific sense, we can create a system that reflects the sensitive nature of the issues and the virtues of a democratic policy. Third, nuclear power requires a process approach because of its peculiar place in our culture. The joint government-industry venture has created a configuration that is neither purely public and democratic nor purely private and capitalistic. Decisions regarding nuclear power were never made exclusively in the free market. Nor were they made as a matter of pure political preference. The public-private combination requires a process through which public policy preferences and normative value choices can be made. A key element to a more responsive and responsible regulatory system is democratic participation. Participation, given low priority in the current structure, is consistent with resolving polycentric problems for which there are no fixed answers, because there is a vast gray area between purely technological or scientific issues and purely legal or political questions. While elite experts may help illuminate the techno-scientific data, they are not better equipped to assess the values inherent in public policy choices.

Having identified the primary characteristics of the transitional period as a focus on cost spreading and decentralization, the regulatory response should embrace a solution to the current problem and create a regime for future policy making that reflects these elements. The themes of the new regime are the traditional values of responsibility and participation. The failure of the nuclear program has been essentially a failure of the primary actors to accept responsibility for their decisions. Correlatively, the persons most affected by commercial nuclear power programs have little say in the decision-making process and are targeted to absorb the costs of these decisions. The failure to allow an avenue of participation is a failure of the democratic process that must be corrected. Therefore, the hallmark of the transitional period is the search for a more responsive regulatory regime.

Having described the postmodern era of nuclear development as a period of retrenchment, a retrenchment coming from a growing public concern about safety, from an increasingly skeptical financial community, and from a legal system reflective of those concerns, if not acutely responsive to them, the remainder of this book sets out the lessons to be learned from the transition.

2. Three Failed Decisions

There is no single villain, nor is there a single victim, in this drama of unfulfilled financial expectations. Even though the action follows a single storyline of industrial and governmental hubris, the consequences of the denouement are many. The image behind this tragedy is much like the smashing of a nucleus. The strong promotional attitude binding the government-industry partnership weakened, and the relationship broke into several pieces. The causes and consequences of the nuclear industry's current predicament will be illustrated with case studies of three nuclear plants: the Zimmer Nuclear Power Station in Moscow, Ohio; the Marble Hill plant in Madison, Indiana; and the Shoreham Project on Long Island, New York. These plants are victims of the changing political and economic milieu besetting the industry during a period of policy adjustment. Each plant has responded differently to these external stimuli: the Marble Hill plant has been canceled and will not be operational in the foreseeable future; the Zimmer plant is slated to be converted to a coal-burning facility; and, finally, Shoreham, although completed, continues to suffer cost overruns from extensive delays.

Each case study presents a brief history of the project, the causes of delay, cancellation, or conversion, and the costs associated with mistaken investment decisions. The case studies establish the factual predicate for the abandonment-costs problem. More important, the studies reveal the need for a coherent regulatory theory on which to base cost-allocation decisions. Later chapters argue that the regulatory system has failed to develop a coherent cost-allocation theory and has not responded to the problem smoothly. This inability to formulate a uniform nuclear policy is disquieting for future regulation.

Since TMI, commercial nuclear policy has undergone radical reevaluation by both market and political sectors. The near-universal nuclear commitment that once existed exists no longer, because the assumptions upon which that commitment was based have evaporated. Nuclear power is no longer the bright hope of our energy future. Rather, it is a tarnished technology that has drained billions of dollars from our economy and has raised safety concerns as never before. Regulators must first attempt to reconcile conflicting interests through an acceptable political process and then utilize the legal system to carry out their choices. There is a pressing need for building a theory that both allocates abandonment costs and forms the foundation for future regulation of the nuclear industry during the adjustment period.[1] The following steps must be taken in constructing an effective regulatory strategy. First, the contemporary commercial nuclear policy must be articulated, capturing the market and political dimensions of the current costs problem. Second, the policy must coincide with the legal rules, methods, and institutions responsible for implementation. Third, the current policy must be adaptable to future needs. Fourth, a framework for reform must be constructed. The strategy will be developed throughout the book and articulated in chapter seven.

Zimmer

PROJECT HISTORY

"We can't afford not doing the best job we possibly can," stated William Dickhoner, president of Cincinnati Gas and Electric (CG&E), the managing and principal partner of the Zimmer Nuclear Power Plant Project on December 2, 1979.[2] That statement was an unfortunate prophecy for the utility. The partners in the venture could not do the job without exorbitant cost escalations. Less than five years later, on January 21, 1984, after numerous government investigations, independent reviews and audits, and escalated estimated project costs, the venture partners announced that Zimmer would not be completed as a nuclear power plant. The partners, CG&E, Dayton Power and Light (DP&L), and Columbus and Southern Ohio Electric (C&SOE), also announced that the nearly completed nuclear project would be converted to a coal-fired unit. After having spent $1.7 billion on the nuclear project, another $1.7 billion would be required for the coal conversion in 1991. One of the perverse quirks of utility law and regulation is that a conversion policy could be considered seriously even though additional billions are needed to accomplish the switch. The policy of encouraging capital investment, ingrained in law[3] for over a century, has reached the unfortunate extreme of forcing inefficient overinvestment.

The rationale and cause behind the abandonment of Zimmer as a nuclear-powered plant can be ascertained from the voluminous documentation accumulated during the plant's controversial fifteen-year history. The fifteen-year "regulatory lag" is a major part of the problem.[4] Zimmer's history shows, among other things, that investment assumptions made in one decade, particularly in a volatile economy, can and do change to the point where a plant is no longer necessary or is no longer economically feasible. Zimmer's peculiar problem can be more succinctly stated in terms similar to Congressman Bill Gradison's comment that "[Zimmer] was shaping up to be the second most expensive nuclear power plant in the United States. . . . [It was] a case of a company that had never built a nuclear plant hiring a contractor that had never built one and employees who had never worked on one. It's been an amateur hour from the very beginning."[5]

CG&E announced its intentions to build a nuclear plant in 1968 and predicted that the plant would be on line in 1975.[6] The company selected the following major contractors for the project: Sargent and Lundy Engineers (S&L), later named as defendants in a $360 million lawsuit as the architect/engineers; General Electric (GE), also named in the lawsuit, as the nuclear steam supply system vendor; and Henry J. Kaiser (HJK) as the major construction engineer.[7] Clearing and construction began after the NRC issued a construction permit in October 1972. CG&E, dissatisfied with the performance of HJK, made a fatal mistake by assuming control over construction management in June 1976. Zimmer was an estimated 50 percent complete at the time. Toward the end of the 1970s, the quality of construction and the safety of the plant came under increasing scrutiny, which ended with the NRC's halting construction on November 12, 1982.[8]

Thomas Applegate was hired as an undercover agent by CG&E in December 1979 to investigate time-card padding by Zimmer employees. Applegate discovered more than CG&E had anticipated. Besides extensive allegations of illegal and dangerously negligent activities by plant personnel, such as drunkenness, theft, and black market smuggling, as well as time-card padding, Applegate reported dangerous conditions in the actual construction of the plant. Poorly executed welds in critical pipelines indicated a collapse of quality assurance (QA) practices at the site.[9] CG&E refused to pursue Applegate's safety concerns. Still haunted by safety violations, Applegate contracted the NRC's Washington office. The NRC consented to an investigation only after Applegate contacted the office of John Ahearne, the chairman of the commission.

The NRC's initial investigation of the Applegate allegations, released July 3, 1980, addressed only three of Applegate's claimed safety violations. The first issue concerned defective welds installed in safety-related systems and was dismissed as unsubstantiated. The second violation concerned defective welds in prefabricated piping that was accepted and installed in safety systems. This complaint was partially acknowledged by the NRC. The only other issue considered was an improper flushing operation; pipes were flushed or cleaned for two weeks rather than the allotted six-week period. A manager at Zimmer said there was no compromise of safety by shortening the time allocation.[10]

Unsatisfied, Applegate alleged that the NRC report inadequately confronted safety violations and completely ignored the discovery of criminal activities, threats to the QA structure, and general mismanagement.[11] He then sought the assistance of the Government Accountability Project (GAP), a public-interest organization associated with the Institute of Policy Studies. At GAP's instigation, the NRC conducted another safety evaluation from January to April 1981. Independently, GAP confirmed Applegate's concerns and found more violations of its own. In May 1981, GAP petitioned the NRC for suspension of Zimmer's construction permit. Its petition focused on four main areas: design, welding, sanitation, and employee drug use. GAP discovered employee reports of designs drawn to conform with construction work already done. Design engineers additionally failed to approve design criteria established to select such items as steel structural beams, small-bore pipe hangers, and restraint systems.[12] The failure to correct identified design flaws had serious consequences. An operator mistakenly forced 1,200 pounds of pressure through pipes meant to carry only 300 pounds, ripping the pipe and spraying water on electricians. Had the plant been operating, the electricians would have been soaked with radioactive water.[13]

More than design problems were discovered by GAP. The petition showed that the NRC was aware of construction problems at the plant. NRC noncompliance reports revealed weaknesses at nearly every step of the welding process. The NRC documented that inadequate training of welders, poor equipment handling, shabby welds, and failure to correct previous citations were a standard part of the Zimmer construction process. The sensitivity and precision of nuclear components protect the public against accidents or shutdown and therefore must be carefully monitored. One example of careless construction is that 75 percent of the Zimmer control rods exceeded specifications. If the rods are larger than required, they may fuse with the reactor vessel during a high-temperature accident, resulting in a meltdown.[14]

GAP's petition to the NRC to suspend construction spurred further inquiry into events at Zimmer. In response to GAP's request for a permit cancellation, the NRC continued its investigation and in November ordered CG&E to establish a quality confirmation program (QCP).[15] The purpose of the QCP was to evaluate and oversee the management of the construction process. The fault with Zimmer, not unlike the fault with TMI, was less technological than human. CG&E's failure to exercise adequate managerial oversight and control caused the breakdown in Zimmer's construction. As a result of the violations cited by the NRC in November 1981, civil penalties were assessed. A $50,000 fine was levied for falsifying records. Reports were knowingly voided and deleted from the noncompliance report system logs and mismatched with reports already filed. Heat numbers, drawings, and weld inspection records did not correspond with respective components in eleven pipelines. A second penalty of $50,000 related to the harassment and intimidation of NRC quality control inspectors. Several inspectors were doused with water while performing inspections and were threatened with physical violence if welds were not passed. This harassment was typical of the annoyances during plant examinations.[16]

The two $50,000 fines for poor construction and sanitation practices can be contrasted with the $100,000 fine also assessed for CG&E's negligence in violating its own QA manual. A quality control inspection program was not established to verify the separation of electrical cables routed from the cable-spreading room to the control room, final inspection of welds on 180 cable tray hangers was not executed as required, and, more critically, the QA division never performed an audit of Sargent and Lundy's nonconformance program.

On November 12, 1982, the NRC ordered CG&E to stop construction. If Zimmer and the NRC had ever enjoyed a honeymoon, it was over. Concerns about construction and management continued throughout the year. A provision of the November order supported a separate NRC evaluation of construction quality at the site, which took place between January 24 and March 4, 1983.[17] The NRC evaluation team identified numerous management and oversight problems as a result of hardware assessment that supported previous allegations. Inadequate field welds and discrepancies in structural steel bolted connections and support plates indicated poor quality control. All work done by Reactor Controls, Inc. was termed suspect and required reevaluation. Many original structural steel welds were of poor quality.[18] The report noted the need for an even closer scrutiny of the design control process at Sargent and Lundy, because drawings and specifications were difficult to translate into useful erection or inspection information.

The NRC ordered CG&E to conduct an independent audit of the construction work and an independent review of Zimmer's project management. Torrey Pines Technology conducted an extensive review of CG&E's management of the project, including its QA and quality confirmation (QC) programs, to determine what measures were needed to ensure construction of the plant in conformance with NRC regulations.[19] Torrey Pines criticized CG&E for allowing a total management breakdown regarding quality assurance. The failure of the QA system results in higher costs due to overruns and delays, and safety failures. A breakdown in the quality assurance program becomes an exercise in fingerpointing. The utility faults the architect-engineer, who faults the contractor, who faults the regulator, who faults management.

Torrey Pines concluded that Zimmer could be completed if a number of management changes were made. Subsequently, Bechtel Corporation, the contractor CG&E hired to replace HJK, issued a revised estimate of the costs to complete the plant. In October 1983, Bechtel projected that it would cost another $1.8 billion to finish the 97 percent completed plant. Bechtel also stated that Zimmer would not be on line before 1986.[20] Facing a final price tag of $3.5 billion, the Zimmer partners decided to pursue an alternate course for the project. Dismal financing prospects and uncertain NRC licensing approval of the completed project were cited as the determining factors in the decision to convert to coal use. The decision not to finish Zimmer as a nuclear plant underscores the irony that regardless of safety problems, choosing a nuclear alternative is a financial matter. The irony is compounded by the realization that the NRC became sensitive to the safety problems only after TMI, and at the instigation of a whistleblower supported by a public-interest organization.

Mr. Applegate's allegations spurred much-needed investigation into the quality control of construction at Zimmer. The NRC's get-tough attitude toward Zimmer was as much a reaction to growing public concern with safety after TMI as it was a reception to Applegate's steadfast complaints. This hardline posture was new for the NRC and was something to which neither the commission nor the industry was accustomed.

As it turns out, the NRC reverted to its old ways after a while. The investigations have not solved Zimmer's problems. The Zimmer plant has not opened and will not be a nuclear power plant. Instead, it is slated to become a coal-fired plant that is as expensive as Zimmer and that may be unnecessary.[21] The Zimmer project has gone from one questionable management decision to another.

CAUSE OF CANCELLATION

The publicly announced reason for converting Zimmer from nuclear to coal was that nuclear was not economically viable. Behind this decision lay the more fundamental fact that CG&E did not know how to build a nuclear plant. The company harbored high-technology hopes with a coal-age mentality, which eventually led to Zimmer's end as a nuclear project.

Although quality-related construction problems were first reported in 1976, the NRC did not conduct a full-scale assessment of plant construction and safety until January 1981. The NRC, on previous occasions, had merely sent in one of its Office of Inspection and Enforcement investigators to conduct routine investigations. No substantive action was taken in any of the reported cases prior to 1981. When the adequacy of those prior inquiries was questioned, the NRC took a different approach. CG&E correlatively minimized safety by not assigning responsibility for inspections until compelled to do so by the NRC. The more serious stance toward Zimmer by CG&E and the NRC corresponds with the changing regulatory attitude after TMI and official reports critical of the NRC.[22] The on-site investigation by the NRC in 1981 was a ten-month investigation of the plant and its management, which revealed a widespread breakdown in the Zimmer QA program attributable to CG&E's failure to exercise adequate oversight and control over its prime construction contractor, HJK, to whom CG&E had delegated the duty of establishing and executing a QA program. By relying on HJK, CG&E attempted to extricate itself from responsibility for construction quality. Not only did CG&E rely on HJK for construction, the utility also relied on the assurances of General Electric, the reactor manufacturer. GE allegedly misrepresented the quality of the reactor and failed to test the component properly.[23] The failure to oversee the work of HJK and GE through QA management was fatal to CG&E's project. The NRC said, with hindsight, that many of the construction deficiencies could have been avoided with proper quality controls. QA, however, was not given adequate attention until the TMI accident was assessed as a problem of human, operator error. Management failure is based as much on ineptitude as it is on misplaced reliance. Zimmer was CG&E's first nuclear power project. As new entrants into the high-technology industry, management wishfully relied on more experienced veterans. CG&E was too deep into the game to withdraw once it suspected that the project was experiencing quality problems.

The NRC investigation revealed numerous examples of noncompliance with twelve of eighteen federal criteria for quality assurance. The problems identified by the NRC centered around the ineffectiveness of the management controls implemented by CG&E and its contractors for ensuring the quality of work performed. Curiously, a majority of the tests conducted by the NRC did not note substantial hardware problems at the plant, because the impact of the quality assurance deficiencies on actual construction was not assessed in 1981. However, recognizing that significant construction deficiencies could have resulted from the quality assurance problems identified during the investigation, the NRC required the establishment of a comprehensive QC program devoted to plant safety.

As a consequence of the 1981 investigation, CG&E paid the $200,000 fine, agreed to increase the size and technical expertise of the QA organization, and agreed to conduct a reinspection of 100 percent of the contractors' QC inspections. The quality control program was to address the following areas:

—structural steel
—weld quality
—adequacy of electrical cable separation
—quality of pipe weld material and radiographs
—adequacy of control design changes
—adequacy of past audits

As could be expected, QA failure was manifested in construction deficiencies. Major construction defects were subsequently revealed, including, in pertinent parts: 1) welds made by an unqualified welding procedure; 2) ASME structural weld and welder qualification deficiencies; 3) electrical cable tray and inspection deficiencies; 4) low-grade structural steel utilized in safety-related construction; and 5) poor pipe support installation.[24] The NRC stated that the large number of nonconformance reports (4,200) and the significance of the matters identified corroborated the NRC staff's finding of a significant breakdown in CG&E's quality assurance program. Numerous other quality-related construction problems in all areas of the plant were also cited by the NRC during its continuing investigation in 1982 and 1983. For example, 70 percent of the structural welds were reworked prior to quality reviews of the welds. The NRC stated that this approach indicated a lack of a comprehensive management program, a failure to address rework activities, and an improper assessment of the safety impact of those activities on the Zimmer plant.

As a result of this information, the NRC concluded, in pertinent part, that:[25]

1. The Zimmer facility was constructed without an adequate quality assurance program to govern construction and monitor quality, which resulted in the construction of a facility of indeterminate quality.

2. Substantial efforts were under way to determine the quality of past construction activities, and numerous construction deficiencies were identified such that reanalysis and rework were required to bring the plant into conformance with regulatory standards.

3. Rework of deficiencies identified by QCP was undertaken prior to the completion of other relevant QCP tasks and reviews, which resulted in the potential for additional reworking of the same item if further deficiencies were found, as had been the case, by the quality reviews.

The NRC said that it lacked reasonable assurance that the Zimmer plant was being constructed in accordance with applicable regulations and that management was inadequate to control the project to meet and maintain NRC requirements.

There is a direct translation of these conclusions into higher costs. All rework activity means delay and more expense. Who is responsible for these higher costs? Is the NRC liable because of an earlier less watchful attitude? Is management liable because of faulty oversight? Or, are the contractors liable for poor workmanship? No party is free from blame.

Because of CG&E's poor history of verifying quality and taking corrective action, the commission concluded that all safety-related construction, including rework activities, should be suspended until there is reasonable confirmation that future activities would be appropriately managed. Zimmer's owners were required to:

> 1. Obtain an independent review of its management of the project, including the QA program, and quality verification. . . .
>
> 2. Submit an updated comprehensive plan to verify the quality of construction at Zimmer . . . the updated plan should include an audit to verify the adequacy of construction.[26]

As part of a stop work order, the NRC conducted an independent evaluation, as well, from January 24, 1983 to March 4, 1983, to further assess the hardware quality at Zimmer, and this investigation revealed similar and more pervasive construction problems at the plant.[27] The visible construction-related problems at Zimmer were only symptoms; the root cause stemmed from CG&E's approach and conduct toward management of the facility.

CG&E MANAGEMENT OF ZIMMER

Following the November 24, 1982 suspension, CG&E contracted with Torrey Pines for project review. The objective of their review "was to conduct an independent review of CG&E management of the Zimmer project, including its QA program and Quality Confirmation Program (QCP) to determine the measures needed to ensure that construction of the Zimmer plant can be completed in conformance with NRC regulations and construction permit."[28] Torrey Pines separately examined key characteristics and aspects of the plant's project management and QA programs. In reviewing over 3,200 documents and interviewing approximately 100 people, Torrey Pines assessed the organizational structure, policies, procedures, and QA activities of CG&E. This investigation included CG&E's communications with its contractors. The review as segmented into four periods:

1. Project inception in 1972 to the assumption of construction management responsibility by CG&E in 1976;
2. 1976 to the NRC's Immediate Action Letter (IAL) in April 1981;
3. IAL to the Show Cause Order in November 1982; and
4. Show Cause Order to August 1983.

Owing to Torrey Pines' emphasis on conducting a review of Zimmer's management, no technical evaluation or review of the plant's design and construction was included. No physical inspection of the plant was performed. Torrey Pines' narrow focus naturally confirmed the NRC criticism of management and steered analysis away from GE and the architects and engineers. Accordingly, the following factors can be identified as the generic causes of the quality-related problems at the plant.

1. *CG&E and Its Contractor Lacked Prior Nuclear Experience.* CG&E was a novice in the field of nuclear power plants, simply having never built one. Further, its prime construction contractor, Henry J. Kaiser Company, had very limited experience in the nuclear field and completely lacked experience as a lead contractor on a state-of-the-art nuclear project. HJK had built only two small noncommercial nuclear plants for the federal government and had not built a commercial nuclear plant prior to Zimmer. Neither CG&E nor HJK had the experience needed to respond in a timely and effective manner to the rapidly evolving, more stringent interpretations of NRC requirements in the late 1970s and 1980s. Rather, they had to follow the lead of the more experienced veteran GE. CG&E did not recognize until

too late that a more formalized, rigorous approach was necessary to control and document the quality of design and construction of a nuclear plant than was required for the design and construction of a fossil fuel plant. Torrey Pines concluded that CG&E's outdated management style was the single most significant factor that contributed to the problems at Zimmer.[29]

CG&E tried to adopt a project management approach at Zimmer that had been used successfully at its fossil fuel plants. The *Owners Project Procedures Manual* established by CG&E for Zimmer in 1972 was merely a modified version of a document prepared for CG&E's Beckford fossil fuel power plant. Such an approach did not honor the complexity of nuclear technology. Significantly, CG&E failed to establish a separate internal organization dedicated to the proper design and construction of a nuclear plant. All key CG&E managers had substantial responsibilities other than the Zimmer project. CG&E's initial company policy for the construction of Zimmer was to: 1) employ reputable contractors; 2) delegate full responsibility to design, construct, inspect, and test the facility; and 3) hold contractors responsible for furnishing an acceptable facility, thus extracting themselves from the process. Because of CG&E's and the prime contractor's lack of experience in the nuclear field, this approach aggravated the plant's problems.

2. *Staff Was Inadequate in Number and Experience to Properly Manage a Nuclear Plant Construction Project.* In comparison with other utility companies, staffing of both CG&E and subcontractor organizations was inadequate throughout the 1970s. CG&E initially had only one or two people at the construction site. Furthermore, CG&E had only one licensing engineer from 1973 to 1979 at Zimmer, who spent approximately 25 percent of his time on other projects. Either HJK requests for additional staff were refused, or the number requested was revised downward by CG&E. Not until after the NRC started scrutinizing the Zimmer project in 1981 did CG&E increase the size of the staff substantially, particularly in the area of quality assurance.

The CG&E staff was found to be lacking in essential qualifications and experience, as well as of insufficient number. Individuals in the nuclear engineering department were hired directly upon graduation from college. Because of CG&E's frugality, nuclear and quality assurance training programs were nonexistent. As Torrey Pines concluded, "the level and status of the CG&E QA organization through the years was generally not adequate to provide for an effective QA program. The major shortcoming in this area was the small and inexperienced CG&E QA staff."[30]

3. *CG&E Failed to Properly Manage the Project.* Several of the key CG&E managers had conflicting responsibilities that detracted from their management of Zimmer. Except for short periods of time, the CG&E manager responsible for the entire Zimmer project was not at the site. CG&E relied extensively on contractors for project management and control. Such reliance is clearly misplaced. The self-interest of GE is to sell

reactors, not construct plants. HJK's primary interest was to construct a plant at its highest level of profits, not at the highest level of safety. CG&E's delegation of authority was inappropriate when the company did not have the accompanying management system, implementing procedures, and staff required to control the work performed by the contractors. The lack of adequate management continued when CG&E assumed control over construction in 1976. The Torrey Pines report states in pertinent part:

> The departments tended to function autonomously at the site. This situation was exacerbated by having the Vice President of Engineering Services located at CG&E's main office, and the department managers, who reported to him, located at the construction site about 30 miles away.[31]

Consequently, the day-to-day awareness of managing and coordinating the project, assigning priorities, resolving conflicts, and solving problems between departments was impaired. Zimmer's construction was guided literally by managerial strategies formulated during the coal age. As a further example, the manager of the General Construction Department at Zimmer was also responsible for CG&E's fossil fuel construction at two other locations. These other projects accounted for approximately half of his time. Torrey Pines concluded that because of the organizational structure established, CG&E's top management lacked a sufficient degree of involvement in, and commitment toward, quality assurance at Zimmer. The president of CG&E did not appear to have an accurate picture of the status and inadequacies of the Zimmer QA program until the Immediate Action Letter of 1981. Prior to that time, the limited executive summaries provided to management did not sufficiently summarize problems or discuss causes and appropriate corrective action. A casual attitude toward managing a nuclear plant is consistent with the state of the industry until the late 1970s. The promotional collaboration between government and industry encouraged both to move projects along to completion as rapidly as possible—that is, until public awareness of safety problems incompatible with a promotional policy became widespread.

4. *Improper Document Control*. Another example of mismanagement cited by Torrey Pines was the lack of document control for reports and records. From the beginning of the project through 1981, there was no single repository for all Zimmer project documents. At least nine independent record control centers were utilized at the time of the Torrey Pines report. This problem contributed significantly to the prevalence and lack of detection of false records and missing documents, documents that were crucial for component and structural verification at the Zimmer plant. Torrey Pines concluded that CG&E's simplified project management system, composed of small staffs and significant delegation of work and overview to contractors, was sufficient for fossil fuel plant facilities, but certainly not for nuclear construction.[32]

5. *Management Policies toward Quality Assurance*. It was apparent that CG&E had a corporate fiscal policy minimizing expenditures.[33] This policy generally benefited both ratepayers and stockholders. However, as Torrey Pines ascertained, the emphasis on minimizing expenditures completely dominated other priorities, such as quality construction and quality assurance. According to the report, cost reduction and schedule maintenance were encouraged to such an extent that contractors and workers attempted to comply only with minimum NRC standards and even failed to attain that level. CG&E did not realize the importance of quality as one of the key elements in the successful construction of a nuclear power plant. This point was manifested in the paltry investment in QA staff. The lack of experienced and qualified personnel on the staff was a factor in the failure to recognize the seriousness of the problems and resolve them in a timely manner. The AEC pointed out these staffing inadequacies to CG&E as early as 1971. However, effective steps were not taken, and as late as 1976, when the project was half completed, CG&E's major priority was to complete the Zimmer project at the least cost and as close to schedule as possible. From this perspective, QA was a requirement to be met at a very minimal level. As a result, there were significant limitations on the effectiveness of the QA program. Torrey Pines revealed that CG&E's QA reviews and audits of contractors for adequacy, implementation, and effectiveness did not consist of the depth, follow-up, or implementation necessary for timely corrective action.

Torrey Pines found that CG&E's president was not aware of the serious nature of the quality problems at Zimmer until 1981, when quality status reports were transmitted to him. They also found little evidence that the CG&E president was actively involved in QA activities. Because of the organizational setup, quality assurance problems simply were not visible to top management, according to Torrey Pines. Further, few audits or special evaluations were requested by upper management.

6. *CG&E Management of QA Program*. As a corollary to CG&E's generally negative attitude toward QA, management of the QA program was deficient, as evidenced in both the original program design and its subsequent implementation. Torrey Pines found that "from the beginning of construction to 1976 there is little evidence that formal training in nuclear QA practices occurred, contrary to the requirements [of NRC regulations]."[34] For instance, project organization was established by CG&E delineating reporting lines for major subcontractors, and defining responsibilities of various positions. Torrey Pines found the plan adequate but not implemented. Further, CG&E did not have an integrated, comprehensive set of project management procedures documented and implemented to ensure that major elements of project construction, engineering, quality assurance, licensing, and scheduling were coordinated. Several instances of inadequate control over design documents, design changes, welding forms, inspection methods, documentation of work accomplished, and conformance to proper

work procedures were also discovered. It appeared to Torrey Pines that CG&E's management was unprepared for the problems of organizing and administering a large nuclear construction project.

7. *CG&E Failed to Respond to NRC Investigations.* During the early stages of construction, many quality problems existed that remained uncorrected because of a lack of attention and follow-through of corrective action.[35] Torrey Pines felt that CG&E was not aggressive enough in pursuing QA problems. Furthermore, rigorous NRC inspection and enforcement did not start until after TMI. Earlier NRC investigations failed to properly document results and failed to determine the correct status and history of several welds. The NRC's Office of Inspector and Auditor concluded that inspections and reports did not accurately reflect the status of the situation at Zimmer. Torrey Pines concluded that CG&E was lulled into a false sense of satisfactory performance by the NRC until the late 1970s and early 1980s.[36]

In summary, it is apparent that the construction problems at Zimmer were indicative of seriously deficient management perceptions and policies of CG&E, compounded by a casual approach toward inspections by the NRC. Minimal costs and rapid completion were valued highly at the expense of quality assurance. CG&E's lack of expertise in the nuclear field, coupled with its delegation of control and oversight to prime construction contractors with insufficient prior nuclear construction experience, further contributed to the ultimate demise of Zimmer as a nuclear facility. Although Torrey Pines indicated in its report that CG&E had taken significant and substantial steps to improve both management policies and quality assurance programs after November 1982, such improvements turned out to be too little, too late. CG&E and its partners, faced with a new cost estimate in excess of $1.5 billion for plant completion, determined that the most viable alternative was to cease construction as a nuclear plant and to convert to a fossil fuel-generated facility.

COSTS ASSOCIATED WITH ZIMMER

In 1968, when Zimmer was first proposed as a nuclear-powered facility, its estimated cost was $240 million. During construction and as delays mounted, the estimated costs rose precipitously. Completion costs at the time the NRC ordered safety-related construction to be stopped in November 1982 were targeted at $1.7 billion. Then, in October 1983, Bechtel Power Corporation, the firm selected by CG&E to replace Zimmer's prime construction contractor, HJK, projected that it would cost another $1.8 billion, bringing the total to $3.5 billion to complete the plant as a nuclear facility. According to CG&E officials, this new cost projection and the associated problem of financing were the prime factors in the abandonment of nuclear plans for Zimmer. In addition, CG&E could not determine with certainty that the NRC would license the plant upon completion. Approximately one week prior to CG&E's announcement that the facility would be converted to coal, the NRC denied a license for Commonwealth Edison's completed nuclear plant in Rockford, Illinois.

When CG&E and its partners in the Zimmer project announced that they intended to pursue an alternate course and convert the 97 percent completed nuclear facility to a coal-fired plant, officials estimated that such a plant would cost about $3.4 billion.[37] Every day that the Zimmer plant sits unfinished along the Ohio River, it becomes $500,000 more expensive.[38] The expenses stem from interest costs accumulating on the $1.7 billion investment. The crucial issue is measuring how much of the Zimmer cost will be passed onto the ratepayer.

Ohio is one of the few states that do not allow recovery from the ratepayers. The Ohio Supreme Court ruled in 1981 that a utility cannot charge customers for the construction of nuclear power plants that it later does not finish and operate.[39] The court reversed a ruling by the Public Utility Commission that permitted the Cleveland Electrical Illuminating Company to amortize expenditures for nuclear plant construction over a ten-year period. The court stated that under Ohio statutory law, a utility may not recover the cost of an investment that never provided any service to the utility's customers. The court commented that if the situation is as "perilous" for utility companies as suggested, the utilities should petition the state legislature to enact changes in the ratemaking structure. The court further noted, in pertinent part:

> Absent such explicit statutory authorization, however, the commission may not benefit the [utility] investors by guaranteeing the full return of capital at the expense of the ratepayer. Under the ratemaking now in effect consumers are not chargeable for utility investments and expenditures that are neither included in the rate base nor properly categorized as costs.[40]

According to this ruling, the significance of which is explored in chapter five, the Zimmer partners cannot recover their investment costs unless and until the plant generates electricity, which thus encourages conversion to coal. Otherwise CG&E is saddled with a $1.7 billion loss. If Zimmer is converted, it is not clear how much, if any, of the Zimmer construction costs associated with the nuclear project can be passed on to the ratepayers. According to a study by the Ohio Consumers' Counsel, only 20 to 40 percent of the existing plant is directly convertible to a coal-fired plant, which suggests that 60 to 80 percent of the investment to date should not be passed on to the consumers. A study by the Ohio Public Utilities Commission attributes 100 percent of Zimmer's costs to mismanagement.[41] Prior to the court decision cited above, the three Zimmer partners had collected approximately $185 million from ratepayers for construction work in progress, which had to be refunded to consumers. Not surprisingly, CG&E president

Dickhoner expected the utility's customers to pay for the new unit; "we see no reason why this should not go into the rate base."[42]

The Ohio case demonstrates how widespread the political and economic issues are. The state agency charged by law to regulate the industry was willing to save industry by passing abandonment costs on to ratepayers. The judiciary stopped it. The executive branch, the governor, argued that the matter was proper for legislative action. All three branches of state government have addressed abandonment costs, the problems have federal ramifications, as well. The quality assurance deficiencies found at Zimmer prompted an even more thorough investigation of nuclear plant management. Congress requested the NRC to assess nuclear plant construction QA nationwide, and the NRC found pervasive management deficiencies, thus implicating the federal government.[43]

Marble Hill

PROJECT HISTORY

In 1973, Public Service Company of Indiana (PSI) proposed a twin nuclear operating facility that would produce 2,260 megawatts of electricity. The original cost estimate was $1.4 billion for both units. Unit No. 1's completion date was targeted as 1982, while Unit No. 2 was scheduled to be on line in 1984.

PSI announced project cancellation on January 13, 1984, with 50 percent of Unit No. 1 and 35 percent of Unit No. 2 completed, and an estimated $2.8 billion invested. The revised projected cost estimates ranged from $7.1 to $7.7 billion for completion sometime between 1988 and 1990 for both units. With most of Marble Hill construction and an additional $4.5 billion ahead, PSI determined that it could not afford to finish the plant. PSI was also of the opinion that additional generating capacity was unnecessary until 1993 at the earliest. Additionally, the Indiana Governor's Task Force, reviewing the Marble Hill facility, noted that the cost savings to ratepayers from nuclear power would not be realized until 1996 or beyond.[44]

The Marble Hill project also experienced quality-related construction problems similar to those of the Zimmer project. Approximately one year after the construction permit was issued, the NRC identified severe deficiencies in the quality of concrete work.[45] There had been several nonconformance reports issued pertaining to poor concrete work from the beginning of the project. Further, the National Board of Boiler and Pressure Vessel Inspectors found code compliance problems with certain piping installations.

The NRC made an extensive inspection of the site, which resulted in the shutdown of all safety-related construction activities. The NRC identified programmatic problems in the PSI project management and quality assurance programs that needed to be resolved before construction could continue.[46]

After an independent review of PSI's management and an eighteen-month delay, limited construction work was permitted by the NRC. A full two-and-a-half years had passed before all construction activity resumed after the issuance of the NRC's stop work order. The costs of completing the plant at that time had already surpassed the original estimates. Arthur D. Little, Inc., a consultant hired for the Governor's Task Force, stated that Marble Hill was not needed to meet the region's energy needs.[47]

As construction delays mounted at Marble Hill, so did ballooning finance costs. At the time of this decision, Indiana allowed a utility to pass on to customers only some costs of canceled plants.[48] PSI, therefore, was forced to spend internal funds and sell more bonds to finance Marble Hill. When new cost estimates were unveiled in September 1983, Governor Orr of Indiana formed a special task force to evaluate whether, with the current cost projections and demand estimates, Marble Hill should be completed. The task force recommended, on December 21, 1983, that the facility not be completed.[49] The task force concluded that the total costs of Marble Hill, estimated at $7.7 billion, did not justify completion when compared with the alternative of supplying future demand with new coal-fired plants.[50]

Since the time of the announced abandonment of Marble Hill, PSI requested a $105 million emergency rate increase from the Indiana Public Service Commission to avoid bankruptcy.[51] Further, one of PSI's partners in the project, the Wabash Valley Power Association (WVPA), a consortium of twenty-four rural electrical utilities, filed suit against PSI for misrepresenting problems with structural steel and quality control during construction at Marble Hill.[52] The WVPA, concerned with its own financial survival, has also proposed converting the facility to a coal-fired unit.

CAUSE OF CANCELLATION

The problems that beset Marble Hill as PSI initiated construction are similar in nature, if not in scope, to those of the Zimmer facility. An individual case study of the Marble Hill project was performed by Pacific Northwest Laboratory in 1983 under the direction of the NRC. Its findings, excerpted here, were published in March 1984 in the NRC's report to Congress.

The study team found that the principal underlying cause of the construction quality problems at Marble Hill concerned poor utility and project management. According to the study, the following "root causes" were significant in contributing to the major quality failures at Marble Hill:[53]

1. *PSI's Inexperience in the Nuclear Field.* PSI had successfully constructed several fossil fuel projects prior to its undertaking Marble Hill. PSI chose to utilize these past management approaches to the nuclear project. PSI's lack of experience and understanding of the complexity of nuclear construction requirements translated into poor management decisions. According to utility economist Charles Komanoff, past nuclear experience

translates into 7 percent lower costs times the number of previously built units. Staffing was simply inadequate in terms of prior nuclear experience, overall qualifications, and size. Although the contractors selected by PSI had fossil fuel plant construction experience, their nuclear experience was depicted as "very limited." Further, PSI delegated extensive overview and management control of the project to the very same contractors. In a role similar to that of CG&E in the Zimmer case, PSI took a back seat to the construction operations at Marble Hill, as demonstrated by the company's minimal presence on site. The study also concluded that PSI did not appreciate the importance of adhering to standards and regulations, primarily the NRC's and ASME's. Finally, PSI was not aware that the recurring concrete and piping quality problems were outward indications of severe programmatic deficiencies in the management of the project.

2. *Quality Assurance*. Senior management officials at PSI were skeptical of the necessity of a formal QA program. Their fossil fuel plants were successfully constructed without such a program. PSI initially established a QA program barely meeting the minimum NRC requirements. The QA staff was inadequate, and when the NRC and the QA staff requested additional support, such requests were ignored.

As a result of this lack of attention to quality, construction integrity suffered. The study team found that PSI believed the contractors when they indicated that the project had no major problems and that similar concrete problems were common in nuclear construction. PSI did not realize the importance that regulatory changes had on the initial concept and design of the plant. Since Marble Hill was similar to many plants being built around the country, the utility owners did not appreciate the significance and complexity of the alterations necessary to complete their plant. PSI acted as a general contractor and construction manager but managed the project from its corporate headquarters. Accountability for the project was delegated among several organizations.

3. *NRC Licensing and Inspection Deficiencies*. The NRC, in its licensing and permit process, did not evaluate whether PSI had the experience, knowledge, staffing, or ability to manage and effectively complete a nuclear plant construction project. The NRC's inspections at Marble Hill were described as erratic by the study team. A resident inspector was not assigned to the site until four months after the stop work order. The NRC, prior to that time, was dilatory in analyzing the comments and evaluations from individual inspectors indicative of the severe generic quality assurance problems in the industry during the 1970s.

It is apparent that the cause of the Marble Hill cancellation was insufficient management control and quality assurance, and, in part, also the NRC's failure aggressively to oversee and expeditiously correct a number of quality problems at Marble Hill. Management failure was endemic throughout the entire project, from the decision to build what is most likely an unnecessary plant through the construction process. Coupled with managerial neglect was management's willingness to blame the NRC's changing regulations or the economic downturn for their failure.

COSTS ASSOCIATED WITH CANCELLATION

Approximately $2.8 billion was spent on the Marble Hill project by PSI and the WVPA. The majority of the project's capital costs were financed through debt and continue to grow as interest accrues. The Governor's Task Force outlined in broad strokes who should absorb the losses of this investment:

> [T]he shareholders of the Company, who stood to benefit from the successful completion of the plant, should bear a substantial part of the cost of not completing the plant. The Task Force also recognized that had the plant's original cost and completion dates been realized, the ratepayers would have also benefitted [*sic*]. Nevertheless, the concept of the ratepayers bearing the substantial burden of the risk of new investments was not acceptable to the Task Force. However, the Task Force believes that it is in the ratepayer's interest to treat PSI's financial situation in a way that will allow the Company to remain viable and to provide access to capital markets when capital is required. Rate treatment alone, however, cannot guarantee PSI's viability.[54]

In a previous nuclear plant cancellation case,[55] the Indiana Public Service Commission allowed a partial recovery of a canceled nuclear plant through a fifteen-year amortization period. The commission favored an equitable sharing of the costs between the ratepayers and common stockholders. It felt the abandonment costs were a legal cost-of-service item. The commission granted a longer amortization period because it would ease the taxpayers' burden. Today, Indiana, by judicial decision, unlike a majority of the states, does not permit utilities to recover abandonment losses from ratepayers.

The Governor's Task Force on PSI made more specific recommendations regarding the cost allocation of the Marble Hill plant. Although the task force concluded that the shareholders of PSI should bear a substantial part of the cost of not completing the plant, it also opined that PSI and its shareholders should not be required to absorb the entire cost if such treatment would impair PSI's access to capital markets, resulting in even higher rate increases. The task force stated that it was a "ratepayer" interest to treat PSI's financial situation in a way that would allow PSI to remain viable.

The task force recommended that if the ratepayer is required to pay in part, then Marble Hill's total cost should be amortized over a twenty-year period. Further, any rate increases should be phased in over a five-year period. The recommendation also included preclusion of dividends for a three-year period and recovery on investment in Marble Hill.[56] The task force stated that if these recommendations are followed, the rate increases for PSI consumers would be less than 3 percent per year. This scheme of rough justice is an attempt to preserve the corporate viability of the utility. The more difficult question is whether the utility, in its present form, should be bailed out. Balancing costs between ratepayers and shareholders has the immediate effect of accommodating the utility. The long-term effects of salvaging utilities with arrested managerial development are questionable.

The task force recognized a different problem, however, in ascertaining a proper allocation of costs incurred by the WVPA. The Rural Electrician Administration (REA) requires the WVPA to pay debt service over a thirty-year period on the money borrowed to finance its share of Marble Hill. The WVPA has no shareholders, so all costs would immediately be passed on to the ratepayers. The task force recommended that the REA mitigate the initial impact of the Marble Hill cancellation by allowing it to phase in rate increases over a period of time and pay the REA accordingly. The task force's proposal was not accepted. On April 14, 1984, the WVPA's $12 million quarterly interest payment to the REA was due. Because the WVPA was unable to pay, the REA ordered the utility association to seek a 51 percent rate increase. The REA wants the entire federal loan repaid.

As a result of this financial quandary, the WVPA proposed converting the facility to coal to help offset some of the debt. The switch from nuclear to coal would take about four years to complete and would cost $2.5 billion. As with the Zimmer conversion proposal, the rationale behind dumping good money after bad is to recoup these expenditures from the ratepayers rather than have them visited on the owners. Governor Orr did not fully endorse the proposal, because the potential energy generated from Marble Hill is not needed in Indiana until the mid-1990s. PSI has stated that it does not have any money to invest in the coal conversion plan. The utility, badly strapped for cash, is spending $3 million a month merely to maintain the twin-silo site. It is expected that PSI will file for a massive rate increase to recover the $2.3 billion it has already sunk into the project.

It is uncertain whether the facility will ever be converted to a coal plant. The only certainty is that it will not be completed as a nuclear plant and that the price in any event for the high-tech monument at Marble Hill is steep.

Shoreham Nuclear Power Station

PROJECT HISTORY

The Long Island Lighting Company (LILCO) ordered the first of its four planned nuclear power plants on Long Island in 1967. The original site was at Lloyd's Harbor on the Long Island Sound. Immediately, the Lloyd Harbor Study Group was formed. With substantial funding from wealthy estate owners in the area, the study group was able to mount a successful campaign to stop LILCO's first attempt to build a nuclear reactor on Long Island. LILCO then chose Shoreham as the new site for the reactor. Shoreham is within the town of Brookhaven on the Long Island Sound. Again opposition, including the Cloud Harbor Study Group, attacked the proposed nuclear reactor. Concern about evacuation in the case of an emergency was voiced in the 1970 Shoreham construction permit proceedings. Evacuation concerns, the AEC stated, should not be considered until the final operating licenses were issued. The attempt to stop the construction failed, and the construction permit was finally granted in 1973. Issues of evacuation and emergency preparedness would come back to haunt and threaten to bankrupt LILCO.

The original estimated cost of the plant was $265 million in 1967. The plant was to be completed by 1975. It was the first plant to use the General Electric Mark II reactor and containment concept, which has become controversial.[57] In 1968, LILCO decided to increase the reactor's capacity from 540 to 820 megawatts without increasing the size of the containment building itself. This maneuver proved troublesome. Site preparation began in 1969, but full construction did not begin until 1973 because of delays in acquiring the necessary construction permit from the AEC.

By the end of 1983, LILCO was still in the process of getting the Shoreham reactor on line. It had abandoned efforts to build other reactors and recently withdrew from the Nine-Mile Point project, a nuclear reactor in upstate New York in which it was a partial owner. In other words, LILCO was not as sure about nuclear power as it had been in the late 1960s. Two points were made by the president of LILCO, Wilfred Uhl, before the 1983 Governor's Fact-Finding Committee: 1) "The plant we built was not the plant we set out to build"; and 2) "If we had it to do again, we wouldn't."[58] These comments refer to the continued change in the design and development of the Shoreham plant and to the extremely high cost that these changes caused. The estimated price of the plant if it begins operation soon is $4.2 billion. However, for each day it is not operating, $1.5 million is added in interest and finance charges. These financial problems may claim another, unanticipated set of victims—the utility employees. In an effort to shore up LILCO because of Shoreham's fiscal drain, the utility hired a new chief executive officer, Dr. William J. Catacosinos. Dr. Catacosinos's internal reorganization efforts were increasingly geared toward averting bankruptcy. The Shoreham plant, at fifteen times initial cost estimates, threatened to ruin the financial viability of LILCO.[59] The real bite came when LILCO sought to reduce employee salaries and benefits in an effort to save the utility. This request was met by a strike of unionized employees.[60]

CAUSES OF PROJECT DELAY

The design and development changes began early in the plant's history. The original plans called for a small reactor, which was increased in size to 820 megawatts, resulting in an unfolding series of expensive design and construction modifications unanticipated by LILCO. The reactor itself was a new design and was only a concept when first ordered. Problems developed, and the design had to be changed. The Mark II reactor was the target of criticism by General Electric's own nuclear engineers, who testified before Congress on February 18, 1976 regarding the safety and design inadequacies of the reactor. The construction contractor, Stone and Webster, was terminated in 1977 after the cost of the project grew and timetable after timetable failed to be satisfied. LILCO had relied almost exclusively on S&W for design and construction modifications, with very little oversight by LILCO. The project under Stone and Webster had become an uncontrolled building experiment by LILCO. Multiple overruns are not a universal element shared by all utilities. Although costly construction problems have been a common experience with the three case studies, some utilities manage to bring in a project near cost.[61]

The issue causing the most delay at Shoreham is inadequate concern for emergency evacuation plans. The limited road network on Long Island, the lack of an off-island road network on the island's east end, the large number of people (over one million) needing evacuation, and the difficulties in ensuring prompt support by adequate numbers of trained emergency workers led to long delays and public demonstrations.[62] Prior to TMI, little attention was directed toward evacuation and emergency preparedness. After TMI, the regulatory establishment underwent a complete reversal. Today, the NRC has extensive regulations pertaining to on-site emergencies, and the Federal Emergency Management Agency controls off-site plans. Even this bifurcated system gets more complicated and overloaded when local governments, in Shoreham's case Suffolk County, fail to participate cooperatively in emergency planning. Each day's delay raises the cost of the project. The evacuation problem has yet to be resolved, and the Governor Cuomo's Panel on the Shoreham Nuclear Power Facility, in December of 1983, after reviewing a number of emergency evacuation plans, including LILCO's and Suffolk County's, decided that there was too much uncertainty on this point to make a recommendation to the governor. In addition, the emergency diesel generator collapsed after testing, indicating a design problem. Full power testing, therefore, cannot begin until a suitable generator is found.

Because of all these problems, plus an unannounced inspection in January 1983 by the NRC that revealed a number of structural and procedural defects, the Shoreham plant did not start its low-power testing until late 1985.

PROJECT COSTS

The financial consequences of the Shoreham project will affect everyone. LILCO's consumers already have the second-highest electric rates in the country. The debt of LILCO by September of 1983 was an incredible $1.6 billion, with interest on that debt ranging from 8 percent to 18 percent.[63] The Shoreham plant, if it operates, will generate at most one-third of the utility's electricity, but its cost exceeds the book value of LILCO's entire electricity system.[64]

The parties potentially liable to pay for Shoreham, whether it operates or not, are LILCO ratepayers, taxpayers, stockholders, and bondholders. Further, the impact of a large increase in rates would also affect the whole Long Island economy. In mid-1983, LILCO asked for a 56 percent increase in rates. The threat of a severe rate hike caused the Grumman Corporation, the largest employer (20,000 employees) on Long Island, to start looking elsewhere. Governor Cuomo arranged to let Grumman buy cheaper electricity from upstate New York in return for an agreement that Grumman expand on Long Island.[65] The governor may not be able to bargain to keep all business on Long Island. State Senator Alfonse M. D'Amato expressed a fear about rising rates: "My concern is that the cost of electricity, whether Shoreham comes on-line or not, is going to drive high-tech high-energy users off the Island. You do that and you're savaging the very core of the Long Island economy."[66]

The New York PSC recently allowed four New York utilities to collect from ratepayers $121.1 million incurred for the canceled Sterling nuclear plant. According to a Department of Energy study, only the New York PSC has violated the "generally adopted practice" of other state utility boards, which do not grant rate increases for the total amount of a canceled plant. The DOE report further states that the New York PSC allows "a fully compensatory return to be earned, thereby indemnifying the utility investor against absorbing any costs."[67] The PSC's recent decision is being challenged by the New York attorney general.

If Shoreham does go on line, the PSC may include in the rate base all sums prudently expended unless the PSC and/or the courts or the legislature change past practices. How much of the cost was reasonably spent on Shoreham is questionable and is currently being unsuccessfully litigated in a precedent-setting lawsuit brought by the county of Suffolk.[68] One possible solution is for the Power Authority of the State of New York (PASNY) to buy all or part of Shoreham by using tax-exempt bonds and then resell the power to LILCO. PASNY purchased one of Consolidated Edison's plants in Indian Point in 1975, but the cost was one-ninth of the cost of Shoreham.

In the end, the longer it takes to solve Shoreham's problems, the larger they become. Continued indecision in determining the plant's fate will lead to a downgrading of LILCO bonds and, as a result, even higher costs of refinancing and, ultimately, electricity. The Shoreham plant is an incredible financial investment. LILCO has to push for its operation or face bankruptcy, especially if it cannot pass the costs on to the consumer. LILCO's claims of pending bankruptcy have not fallen on deaf ears. The PSC granted LILCO a 9.6 percent rate increase for the express purpose of avoiding bankruptcy.

Paul L. Gioia, the PSC chairman, said that the rate increase was intended to send a signal to the banks that are trying to decide whether to loan LILCO the money it needs to stave off bankruptcy.[69] However, New York will not allow imprudently incurred expenses to be passed through to consumers. On the other side, the plant has been through so many changes that one wonders if it can operate safely. The safety contention along with the inadequacy of evacuation plans raises the problem of negative public opinion of LILCO. Because of the strong possibility that operation will be delayed for at least another two years, it may do LILCO well to cancel the plant or convert it. The daily financial costs alone would argue for this choice. In any event, a financial mistake was made, which, even if the plant operates for any length of time, will cost everyone, including LILCO, a great deal and threaten to make LILCO power the most expensive electricity in the country.

CASE STUDY SUMMATION

The three case studies depict the transitional phase of nuclear power regulation as a period of confusion. Each plant can identify a different cause: Shoreham's evacuation difficulty, Marble Hill's excess capacity, and Zimmer's poor quality. The state regulatory commissions have reacted differently to allocating abandonment costs. Ohio and Indiana do not allow the utility to collect the costs from ratepayers, but New York does.

Even though the case studies reveal different causes and different consequences of the abandonment problem, the studies share a common failing from which a theory can emerge.[70] The primary cause of the significant and costly delays and eventual loss of nuclear faith is poor project management by both utilities and the NRC. This breakdown in nuclear construction is emphasized more as human, managerial failure than as a scientific or technological failure. The construction quality problems prevalent at each of the three nuclear plants were the end result of serious deficiencies in managerial approach and conduct. The deficiencies stemmed from the utilities' lack of prior nuclear experience and their attempt to build a nuclear plant with fossil fuel management principles. Quality defects were also the result of NRC laxity in safety inspection and enforcement. This mistake in managerial judgment is another example of how nuclear technology has outrun the industry's ability to control it. The managerial, as well as the regulatory, response to nuclear power must be correlated to the demands of the new technology. Thus far the response has been wrong. The demands of high technology should not have precluded utilities from successfully completing projects without developing major quality problems. Each utility owner of the approximately eighty nuclear plants now in operation in the U.S. was at some time a first-time owner. It is noteworthy that the first commercial nuclear plant at Shippingport was constructed under the management of people who had extensive prior nuclear design and construction experience in the Navy's nuclear program. Furthermore, a number of the early plants in the United States were "turnkey" plants that were constructed by a few large architectural and engineering firms and nuclear reactor vendors. These firms, whose first reactor plants were far simpler than today's, have developed a base of experience from which they can draw for constructing the increasingly more complex reactors at lower cost.

These three plants, and the dozens that are being canceled, have failed because the poor energy-economic climate has caught up with poor management techniques. In the 1970s, when the Zimmer, Marble Hill, and Shoreham projects were initiated, there was a large block of orders for new plants. The demand for personnel and organizations with successful prior nuclear design and construction experience exceeded that supply. Additionally, competition among the four major reactor vendors (Westinghouse, General Electric, Combustion Engineering, and Babcock and Wilcox) caused each firm to market new, but untested, designs, exacerbating design and manufacturing problems.[71] Consequently, utility owners had to choose key project team members from either the "reserve" members of an experienced firm (personnel lacking sufficient experience) or the "first team" from a firm that was inexperienced in nuclear design and construction.

As evident from the case studies, this supply-and-demand problem for project team members with prior nuclear experience, coupled with the inexperience of the owners themselves and their willingness to let outsiders control construction, led to situations where inappropriate project roles were assumed beyond team members' capabilities. Managers simply assumed or assigned roles in the nuclear project with which they were unfamiliar, and quality problems surfaced.[72] The owners' inexperience contributed to their failure to fully appreciate the complexity of their nuclear power plant projects. As a result, in their rush to get plants on line in the least costly way possible, the utilities merely adopted management practices and project approaches that had been successful in the construction of fossil fuel facilities. This significant lack of differentiation between fossil and nuclear technologies contributed to the demise of the successful completion of nuclear plants. Utility owners failed to implement a management system that would ensure adequate control over all aspects of the project. The deficiencies in the management systems were manifested in the following areas:

—inadequate staffing for the project, in numbers and qualifications;

—contractors with sufficient fossil fuel experience but limited nuclear experience;

—reliance on these same contractors in the management and quality control of the project;

—lack of management commitment to quality control and confirmation;

—lack of appreciation of codes and other standards;

—misunderstanding of the NRC, its role, practices, authority, and requirements;

—inability to recognize that recurring problems in construction quality were symptoms of underlying problems in project management.

Although the utility owners failed to implement appropriate management systems that might have alleviated the majority of the problems that occurred, the NRC, to an extent, did not adequately conduct its role both prior to and during construction. The NRC's neglect is glaringly apparent in the Zimmer case, where well-founded allegations of construction quality problems surfaced years before the NRC conducted a comprehensive investigation. This investigation was prompted only by the persistent allegations of an intervenor whistleblower and former employees of the Zimmer project. The NRC's response was also stimulated by growing public awareness of the safety hazards of nuclear power. Nuclear power regulation has been assaulted from two sides at once. The market demands cost reductions, while political voices demand increased safety inspections.

A recent study conducted by the NRC confirmed serious inadequacies in the commission's program and practices. The study found that the NRC did not properly screen utility owners during the construction permit process to determine whether the utility possessed management capable of undertaking a nuclear project. Further, the study concluded that the NRC had been remiss in its goals of 1) preventing major quality-related problems; 2) detecting, in a timely manner, quality problems and taking appropriate corrective action; and 3) assuring the plants licensed to operate have met applicable legal requirements and are designed and built in a manner consistent with public safety. The study stated that the NRC was not primarily responsible for accomplishing any of the above items, but the NRC, as the architect and monitor of the QA system, must share in the blame when the system does not work.[73]

In its study, the NRC also reviewed the construction management approach at successfully completed and operating plants. It found that the essential characteristic of successful projects was prior nuclear construction experience of the project team.[74] Prior nuclear construction experience of utility owners was also helpful, although not necessarily mandatory, if the rest of the project team was sufficiently experienced and had assumed commensurate responsibilities. The NRC report also stated that utility owners viewing NRC requirements as a minimum, not a maximum, level of performance were more successful in completing their projects.

The necessity for a more self-evaluative and critical managerial attitude indicates how complex the abandonment problem has become in attempting to assign blame and allocate liability. The nuclear power industry marched headlong into commercialization with government encouragement. Both did so with a blind faith that nuclear power would work out and kept their eyes closed and fingers crossed as problems developed. TMI shattered that idyll. After TMI, industry reacted to the exorbitant costs, and government reacted to the public's concern about safety. The partners in the nuclear venture began to see their primary concerns differently, with the same realization that a safe plant was not a cheap investment. Finally, it is imminently clear that neither industry nor government can afford a cavalier approach to safety and quality of nuclear facilities. The case studies demonstrate that finances and safety affect the future viability of the industry. The allocation of abandonment costs must be grounded in a liability theory that takes both a retrospective look at past responsibility for failed decisions and a prospective look at future regulation. Utilities or builders should not be bailed out to make the same mistakes again. Nor should the regulatory system encourage capital investment at the expense of plant safety.

3. Public Participation

The case studies depict the regulatory and construction histories of three canceled nuclear plants. Interspersed throughout the case studies are references to intervenors. The role and impact intervenors have on nuclear and electric utility regulation are difficult, if not impossible, to assess accurately. The assessment is problematic because the place of intervenors in the regulatory state is subject more to rhetoric and polemic than to cool scholarly analysis. According to regulators and regulatees, antinuclear advocates and consumer advocates are the cause of costly delay.[1] According to intervenors, the system is institutionally biased against them.[2] The antipathy toward intervenors by regulator and regulatee is understandable but not defensible. Intervenors upset the apple cart by challenging existing procedures, suggesting alternatives, and questioning basic assumptions. The primary opposition to intervention is that it is inefficient.[3] This assumption is questionable at best. At worst, it is just plain wrong.[4]

Answers to questions about the role of public participation in the regulatory process are both theoretical and practical. First, regulation of utilities requires public participation as a matter of constitutional law and democratic principle.[5] The right to participate in public decision making is not granted simply because some law says the right exists. Constitutional arguments, specifically ones resting on fairness, due process, and access to public decision making, are anything but technical. Instead, the constitutionally guaranteed opportunity to participate ensures the legitimacy of our democratic polity. Politics without participation is autarchy, not democracy. Second, intervenors raise issues that would not be raised otherwise. Sometimes these issues are more costly to utilities but protect the safety of the public. Sometimes issues raised by intervenors save money in the long run for utilities. In either case, affected individuals become actively and usefully involved in policy-making and decision-making processes. Public participation has both economic and noneconomic values. To demonstrate the nature and role of public intervention, the Zimmer nuclear power station will serve, again, as a case study from which broader principles will be drawn.

Behind the rhetoric and polemic lie deeper reasons for the difficulties attached to evaluating the role and impact of intervenors. The basic difficulties concern choice of evaluative measurements and differing frames of reference. If intervenor participation is measured by a cost-benefit yardstick with which delay is equated with costs, then any cost increase is bad. Any delay naturally results in the imposition of at least short-term transaction costs. If intervenor effectiveness is measured in terms of the ventilation of issues not otherwise presented, the presence of voice in decision making, and the occurrence of long-term benefits, then another set of parameters is used. The choice of evaluative test, therefore, is determinative. Further, the tests indicate two dualities present in nuclear regulation. One test attempts to measure short-term and quantifiable effects, the other purports to measure long-term and nonquantifiable effects. These tests are mutually exclusive ways of looking at the world.

The nuclear transition has witnessed increased public participation in nuclear and electric utility regulation. The rise of intervenor participation is directly responsive to market conditions. As electricity costs rise, ratepayers have an incentive to oppose increasingly higher rates. However, individual representation in PUC hearings remains ineffective. Instead, organized intervention is required. Even then, organized efforts must combat a century-old mindset of traditional ratemaking and four decades of nuclear promotion. This chapter presents commentary on certain important facts and inferences derived from those facts assessing the effectiveness of public participation.

A BRIEF HISTORY OF PUBLIC PARTICIPATION

Public participation and nuclear power have never mixed well.[6] In part, that is a result of an elite bureaucratic model of nuclear regulation. Nuclear physics and its implementing technology were beyond the ken of ordinary laymen. Indeed, it was beyond the understanding of everyone but the physicists responsible for the development of commercialization. The governmental response to technological complexity was bureaucratic specialization. The assumed best bureaucratic model was one in which expertise was harnessed to regulate a complex industry.[7]

We can accept certain facts and disagree about the conclusions to derive from those facts. Nuclear science and technology are complex. They are also uncertain. Even though scientific and technological complexity may best be left to experts, uncertainty should not be. In the interstices of uncertainty and complexity are politics and policy making.[8] Further, competing private interests are mediated through public participation. The expert, elitist model of bureaucracy has the undesirable consequence of muffling the public voice.[9] Such has been, and continues to be, the history of federal nuclear regulation. Public participation has been more effective in state PUCs, where the financial consequences of utility regulation are prominent.

During the 1950s, there was little opposition to plans for the construction of nuclear plants. Even then, opposition to AEC proceedings was unwelcome.[10] Opposition has increased as more citizens have become disenchanted with nuclear power.[11] The lack of public confidence stems from the perception that the regulatory agency and the nuclear industry are "a community of vested interests."[12] Historically, government agencies have protected the industry from a public that might react "with an irrational and emotional response that would retard growth of the technology."[13] Within the licensing process, the risk of accidents was characterized as "impossible," "incredible," and "extremely remote." Minimal public participation was used merely to give credibility to the proceedings. According to Professor Harold Green:

> The fourth problem area lies in what is euphemistically characterized as "public participation" which is so tenderly preserved, yet so delicately curbed, in the NRC and DOE reform proposals. Eleven years ago I wrote that public participation is at best a charade and at worst a sham. Nothing has happened in the intervening years to change my view. Public participation gives ordinary citizens and citizen groups the opportunity to retain counsel, employ nuclear experts, and become parties in adversarial, quasi-judicial hearings. The concept of public participation evolved after the fact to characterize and rationalize the phenomenon that numerous intervenors availed themselves of a statutory right to a hearing in licensing proceedings in order to harass, impede, and hopefully obstruct specific nuclear power projects. It was not a case of Congress or AEC/NRC saying "let there be public participation" and then providing adversary hearings to achieve this objective. Quite to the contrary. . . ."[14]

As a consequence, citizens began to view the Atomic Energy Commission and its successor, the Nuclear Regulatory Commission, as captured agencies[15] more committed to the development of the nuclear industry than to its regulation. Certain advocates rejected the "myth that only technical experts [were] capable of forming valid judgments as to scientific and engineering applications in public policy issues. . . ."[16] With increased regularity, opposition groups demonstrated and marched, while in a less dramatic but more significant way, intervenors petitioned their way into technically complicated adjudicatory hearings.

Public concern with nuclear power has resulted in increasing intervention by local interest groups in the licensing hearings for nuclear plants. These intervenors include individual citizens, local and national citizens' groups, environmental organizations, localities, and states. The groups have been divided into two broad classifications by William Gormley as grassroots and proxy advocates.[17] The distinction between the two classes is important. Grassroots organizations, at either state or national levels, are private interest groups. Proxy groups, such as the Ohio Office of Consumers' Counsel, are empowered by the state and are authorized to represent persons, such as consumers, who have been disenfranchised from the regulatory process. Some proponents of the nuclear industry imagine a conspiracy between well-financed anti-nuclear power activists and sympathetic courts. The reality is that more intervenors are grassroots or proxy advocates concerned only with a specific facility,[18] the two most notable exceptions being the Union of Concerned Scientists and the Natural Resources Defense Council. More important, intervenor groups represent a variety of interests and are not monolithic. Some intervenors stress nuclear safety; others stress lower electricity costs. Safety and low-cost electricity are ends that do not coincide.

Courts and commentators have noted that intervenors have made positive contributions in bringing information and issues to the public's attention, as well as in surfacing additional material for the regulatory agency to consider.[19] Public participation enhances confidence in agency proceedings and exposes its hearings to public scrutiny. In the late 1960s and early 1970s, intervention was premised on environmental and ecological issues. Nuclear power plants require sites contiguous to bodies of water and thus compete for limited environmental resources. In the mid to late 1970s, nuclear safety issues gained prominence. The accident at Three Mile Island created even more doubt about the technological uncertainties of nuclear power and less doubt about the possibility of a major accident or a prolonged shutdown. This episode undermined the public acceptability of nuclear power and increased skepticism over the NRC's supervision of the industry. Since TMI, the financial aspects of the industry have occupied center stage.

REGULATORY PROCESSES

To appreciate the role of intervenors, it is necessary to understand their position in the regulatory process.

Although agencies have a wide variety of functions and procedures, they exercise basically two powers: rule making and adjudication. Rule making is prospective and legislative.[20] It is forward-looking and is intended to affect large numbers of unidentified persons. Adjudication, at the other extreme, is retrospective and dispute-resolving. It settles past conflicts between known, identified parties. In both instances, the public is positioned outside the process and must move to be included inside. Regulatory hearings range from formal (adjudicatory) to informal (rule-making), at which parties other than the agency or the utility can present views. In some instances, agencies will accept oral or written comments from any "interested" person. The informality of this form of participation can preclude effectiveness. Formal adjudicatory or trial-type hearings are more onerous because of the expense involved and the expertise necessary for effective participation.

The administrative procedures developed by the Nuclear Regulatory Commission, for example, specifically delineate the steps citizens must follow in order to present issues. Through its two-step construction and operating licensing process, the NRC addresses matters it is required to consider under the Atomic Energy Act[21] and the National Environmental Policy Act.[22] The NRC is charged with overseeing public health and safety,[23] technical and financial qualifications of the applicants,[24] and environmental impacts under the NEPA.[25] At least one year before the construction permit application is made, the utility must initiate a series of discussions with the NRC's regulatory staff. The staff evaluates information provided by the company. After the utility's application is accepted and docketed, the staff becomes a mandatory party to the adjudicatory proceeding. This preapplication review is not a public proceeding, nor are intervenors allowed to participate.

After the utility tenders its application to the NRC, notice of the application is published in the Federal Register. Copies of the application are furnished to affected state and local authorities and to a local public document room near the proposed site, as well as in the NRC's public document room in Washington, D.C. Through this notice procedure, the public is invited to intervene. Potential petitioners must file twenty copies of their petition, listing with specificity the contentions to be addressed and the bases for each contention.[26] At this point, discovery by intervenors can begin.[27] At least twice before the commencement of NRC licensing hearings, prehearing conferences are held. The final prehearing conference results in an order that determines what issues are contested and establishes a hearing schedule. Licensing hearings resemble a trial, in that the hearing is individualized and keyed to an individual project.[28]

Once the licensing board has approved the application, the staff automatically issues a license authorizing the beginning of construction, triggering the expenditures of large amounts of capital.[29] The decision is subject to appeal, and intervenors can halt construction by applying for an order staying the NRC's immediate effectiveness rule. Because of traditional deference to administrative agencies, court reversal of decisions is rare. Intervenors can also appeal the order to the Atomic Safety and Licensing Appeals Board, which has the power to rewrite the ASLB decision. Like its judicial counterpart, it rarely negates the licensing board's decision.[30]

Two or three years before the nuclear facility is scheduled to be completed, the utility files an application for an operating license. A process similar to that for the construction permit is followed. The application is filed, then the NRC staff presents a safety report and an updated environmental statement. Although a public hearing is not mandatory, one may be held if either intervenors or the commission requests it. Once licensed, the plant remains under NRC surveillance for its operating life.

Administrative procedures are lengthy and technically complex, which makes it very difficult for the lay person to absorb or understand the data. Even well-organized public participants must present their case against an institutionalized backdrop that is resistant to change if not hostile to intervention.[31] Notice is inadequate (unless one is a faithful reader of the Federal Register), and the initial meetings between the utility and the NRC are closed. All of these features act as disincentives that deter private citizens from becoming involved. However, the major obstacle to public intervention is money. Few individuals or groups have the financial resources to cover administrative costs, lawyers' fees, or the hiring of technical experts. Although public participation is permitted, administrative regulations and bureaucratic inertia clearly favor the utility industry while discouraging all but the most committed intervenors. In view of this bias, it is difficult to ascertain just how instrumental intervenors are in nuclear policy making.

The participatory process in state PUC hearings is similar. The burden is on intervenors to become aware of issues that may affect them, and they then must move to enter the process. Once inside, either formality or informality may preclude effective participation. Informality can negate the effectiveness of public participation because of its pro forma quality. Similarly, triallike formality can preclude effectiveness because of the potential for delay and the use of legal technicalities, which can stop all but the most prepared participants. The key to effective participation is organization.[32] That is true of grassroots and proxy organizations. In states with active rate counsel, more active and effective cases can be noted.[33] In situations with more organized grassroots groups, effects are also seen.[34]

ZIMMER CASE STUDY

The role and influence of intervenors in the cancellation of a nuclear power plant can be illustrated by focusing on the cancellation of the Zimmer plant. This description is an attempt to trace the efforts of various groups participating in the Zimmer licensing hearings to determine what part they played in the decision-making process. In many respects, the events preceding the Zimmer cancellation had all the elements of a heroic epic. The drama involved dedicated intervenors, government-industrial powers, a national disaster, and a disinterested private investigator turned whistleblower.

In 1968, Cincinnati Gas and Electric Company, with its partners Dayton Power and Light and Columbus Southern Ohio Electric Company, announced plans to construct an 800-megawatt nuclear power plant.[35] Given industry forecasts and national energy planning at the time, the proposal to build seemed reasonable.[36] Following technical reviews by the AEC staff and the Advisory Committee on Reactor Safeguards hearings and a favorable initial decision by the Atomic Safety and Licensing Board, the Atomic Energy Commission issued a construction permit on October 27, 1972. The building of the nuclear plant got under way on a 600-acre site twenty-four miles southeast of Cincinnati. Commercial operation of the facility was projected for 1975. In May 1975, the Cincinnati utility applied for its operating license.

After the NRC docketed the application for an operating license, notice of the application and the opportunity for hearing was published.[37] In response to the notice, petitions to intervene were filed by the Miami Valley Power Project (MVPP), Dr. David Fankhauser, a Clermont College biology professor, Mrs. Marie B. Leigh, and the city of Cincinnati.[38] As required by NRC regulations, the various intervenors raised different claims. The MVPP was primarily concerned with safety-related issues. Dr. Fankhauser's focused on the adequacy of the emergency evacuation plan.[39] At the urging of the Environmental Advisory Counsel, the city of Cincinnati intervened to insure that a proper water-monitoring system was installed.[40] According to John Woliver, legal counsel for Dr. Fankhauser, the intervenors felt that barring some dramatic turnabout, the granting of the license was a foregone conclusion.

TMI caught both the nuclear industry and the NRC unprepared. "What shook the public the most," said Victor Gilinsky of the NRC, "was seeing the men in the white lab coats standing and scratching their heads because they didn't know what to do. The result was that accidents were taken seriously in a way they never had been before."[41] TMI led to investigations, and the NRC listed over six hundred steps that utilities had to take to improve the safety of the plants. As the licensing hearings for Zimmer opened in Cincinnati in June 1979, CG&E, citing changes in the NRC requirements, projected that the plant would now cost $850 million and would not be operable until early 1981. TMI coupled with CG&E's announcement resulted in renewed efforts by intervenors. The city of Cincinnati was now concerned about an early-warning system that would immediately provide the city with information about the concentration, amount, direction, and rate of speed of any released nuclear particles. Lawyers for the city sought to have a sophisticated tracking system with automatic feedback and constant updating installed at the plant. While CG&E would not provide the exact system requested, the company did agree to install an early-warning mechanism. Consequently, in October of 1981, the city withdrew as an intervenor and agreed not to interfere with the operating license proceedings.[42]

In April 1979, Dr. Fankhauser reasserted his claim that the applicants had no plans for or knowledge of the shipping of waste material. CG&E had also made no provision for training people in the community to cope with transportation accidents. Furthermore, the issues of safety transportation were dependent on transportation routes that had not been chosen. Because of TMI, new regulations concerning this issue were not fully developed. The staff advised the licensing board to defer a ruling on the merits.

Two months later, CG&E filed a "Renewed Motion" to have the complaint summarily dismissed. The company asserted that the new rule involved the prevention of sabotage of shipments of radioactive materials and made no provisions for training the communities about the transportation of radioactive materials. In response, Dr. Fankhauser argued that the new rule required the applicant to make plans for the routing of spent fuel and that CG&E had not made any such plans. Two weeks later, the board determined that there was no training requirement of nonirradiated fuel. That left open the question of the transportation of irradiated fuel. The board invited all parties to submit additional comments on the applicant's motion, taking into account that the interim rule would become finalized on July 3, 1980.

The NRC staff, Dr. Fankhauser, Zimmer Area Citizens (ZAC) and Zimmer Area Citizens of Kentucky (ZACK), the MVPP, the city of Mentor, Kentucky, and the Commonwealth of Kentucky filed responses with the board. The intervenors were concerned that security measures should be considered in the proceeding. Because federal regulations require that the licensee only had to prepare a plan for the physical protection of spent fuel against sabotage, the board decided that there was no requirement that this plan be submitted prior to licensing. Intervenors had not shown any unusual circumstances that might suggest that the protection of spent-fuel transportation from either sabotage or accident needed to be considered. Nevertheless, the board suggested that further review, affording an opportunity for public participation, might be warranted.[43]

During the period in which the board was deliberating Dr. Fankhauser's initial claim, ZAC and ZACK petitioned on March 21, 1980 to intervene in the operating license proceeding. ZAC-ZACK was concerned about the health of its members, the safe operation of the facility, "and the effect upon petitioner's safety and property in the event of emergency, particularly in view of its members' homes being located within a ten-mile radius of the facility and its children attending schools within that radius." CG&E opposed the intervention, claiming that the proceeding had commenced more than four years previously and that the petition should have been filed by October 1975.

In response, ZAC-ZACK maintained that it could show "good cause" for the delay. The two organizations had not been formed until after TMI and had not sought to intervene until they had achieved a degree of expertise on the issues. The group also claimed that intervention was appropriate in view of the regulatory revisions mandated by the TMI experience. In fact, off-site emergency preparedness became a part of the licensing process only after TMI.

The board agreed that new regulatory developments in emergency planning did constitute "good cause." Even though Dr. Fankhauser and the city of Cincinnati had both raised these issues in 1975, the emergency planning at the time would have extended a radius of three miles from the facility. After September 1979, the "emergency planning zones" were extended to ten miles from the plant and for ingestion pathways of fifty miles.

ZAC-ZACK could provide the board with practical working knowledge of transportation and traffic conditions relevant to emergency planning. No other intervenors presently involved had or could represent the groups' concerns with respect to the interests of school children residing within ten miles of the facility in both Kentucky and Ohio. Finally, because the NRC's hearings on evacuation and monitoring were at a standstill pending the development of new criteria, ZAC-ZACK's intervention would not cause delay in the proceedings.[44] As a result, the board granted ZAC-ZACK leave to intervene.

The plant's other organized opponent, the MVPP, submitted additional complaints about plant safety. The group was aided by private investigator Thomas Applegate. In November 1979, Mr. Applegate was working on a divorce case involving an employee at the plant. During his probe, he discovered evidence of time-card cheating by Zimmer workers. When he took his results to the utility's management, CG&E hired him to investigate further personnel abuses. Applegate's investigation soon produced evidence of other false documentation, and at this point, utility officials let him go. Concerned that this information, which included theft of materials and evidence of defective welding, was not handled responsively, Applegate took his findings to the NRC. The federal investigation that followed resulted only in a citation by the commission for minor paperwork violations.

Mr. Applegate persisted in his allegations, and in May 1980, he contacted GAP. Following Applegate's leads, GAP investigators found evidence of an inadequate NRC probe and convinced the commission to conduct a second inspection into the quality and safety of the plant's construction.[45] In January 1981, the NRC began a second look into Zimmer, which uncovered violations of twelve of eighteen basic safety criteria for building a nuclear plant. It also revealed that the quality assurance program was too understaffed to check on construction. Inspectors turned up thousands of nonconforming or undocumented welds, as well as allegations from quality control inspectors of harassment and intimidation by plant workers.

As a result, in November 1981, a $200,000 fine, the largest against a nuclear power plant still under construction, was levied against Zimmer. CG&E paid the fine but claimed that the NRC had exaggerated the problems at the plant. Several months later, the company informed its shareholders that the cost of completing Zimmer had risen to $1.5 billion.

In the wake of these developments, the MVPP filed a Motion for Leave to File New Contentions on May 18, 1982.[46]. The motion stated eight new contentions concerning two general areas: the status of quality assurance pertaining to the construction of Zimmer, and the corporate character and competence of CG&E to manage a nuclear generating station. Specifically, the motion charged that CG&E and Kaiser Engineering had failed to maintain sufficient quality assurance controls to ensure that the as-built condition of the plant reflected the final version of a design that complied with federal requirements for public health and safety. Second, the applicants could not provide adequate material traceability to identify and document the history of all materials, parts, components, and welds. The utility did not have a sufficient quality assurance program for vendor purchase. Moreover, CG&E and Kaiser had not corrected construction deficiencies, nor had they made good-faith efforts to comply with the audit recommendations. Company officials, the MVPP alleged, had not maintained enough controls to process and respond to internal nonconformance reports identifying violations of internal governmental requirements. The motion also alleged that CG&E and KEI had engaged in illegal retaliation against quality assurance and quality control personnel who had attempted to disclose QA problems to the NRC. The MVPP claimed that the quality confirmation program could not mitigate or remedy the serious consequences of QA breakdown at Zimmer. Finally, the petition contended that CG&E lacked the necessary character and competence to operate a nuclear power plant.

Even though the Atomic Safety and Licensing Board noted that the MVPP's contentions were untimely, it determined that the issues raised by the allegations not only were very serious but also had the potential for resulting in a possible denial of an operating license. Consequently, on July 15, 1982, the staff urged that the public interest would best be served if the board would reopen the record and litigate the contentions. The public airing of the matter, the board felt, might strengthen public confidence as well as the quality assurance program at Zimmer.

In early June 1982, at a licensing board hearing, ZAC-ZACK and the city of Mentor challenged the adequacy and capability of the various state and local emergency plans submitted by CG&E. The issues involved specifically the emergency planning in Clermont County, Ohio and Campbell, Kentucky to effect a timely evacuation of school children in nineteen elementary and secondary schools in the event of a nuclear accident.[47] Under federal rules, Zimmer officials must have the capability of notifying responsible state and local governmental agencies within fifteen minutes after declaring an emergency.[48] According to the proposed plan, officials at the nuclear facility would communicate news of an emergency to school administrators via commercial telephone.[49] The public in general would be notified through sirens, weather radios, door-to-door verifications, and an emergency broadcast system.

The board found that these plans did not provide for sufficient assurance that key persons would receive prompt notification of any emergency in case an evacuation was necessary. They reasoned that commercial telephone circuits would become overloaded during an emergency, and thus unavailable for official use. The communication problem was compounded because no plans had been developed for mobilizing school buses if telephone service became unavailable. On June 21, 1982, the board issued an initial decision in advance of its ruling on the MVPP's earlier motion in order to expedite any additional proceedings that might be necessitated by its decision. Dr. Fankhauser's contentions were summarily disposed of favorably to CG&E. The MVPP claims questioning the need for the facility, the adequacy of a fuel supply, and the financial qualifications of CG&E, together with several allegations of construction flaws, were also disposed of favorably to the utility. However, the board concluded that the deficiencies in the off-site emergency plans had to be corrected before an operating license would be issued.[50] They refused to authorize the license until the Federal Emergency Management Agency issued findings on the adequacy and implementability of the evacuation plans and the parties had an opportunity to assess the plans. Considering that the board disposed of numerous contentions, it was surprising that ZAC-ZACK managed to forestall the operation of the nuclear plant with one of its issues.

The "victory" would prove important in upcoming months. On July 30, 1982, the NRC agreed with the Atomic Safety and Licensing Board that the issues presented by the MVPP were serious. The commission concluded, however, that the NRC had been investigating alleged quality assurance irregularities at Zimmer since January 1981. This investigation was ongoing, and the eight new contentions raised were repetitious of problems revealed in reports already released to the public. The MVPP's petition not only was untimely, but also it did not contain any new information. By a 3-2 vote, the commission could find no justification to support an order to reopen the hearings.[51]

A month later, GAP petitioned the NRC to stop construction and to reconsider its ban on further hearings into safety problems at Zimmer. GAP contended that the NRC's 1981 report on QA problems had identified only a small number of the deficiencies and had omitted significant affidavits. Furthermore, they charged that the final report had been altered to justify less drastic action against CG&E to answer these allegations by December 31, 1982.

By the end of October, the city Environmental Advisory Council (EAC) recommended that the Cincinnati city council request a reopening of the Zimmer licensing hearing. Because of the prior agreement with CG&E, the city council was reluctant to raise the issue. However, the EAC argued that the city could no longer delegate the responsibility of ensuring safety to other intervenors. Reopened hearings would also increase public confidence and assure the correction of safety problems at the plant.

NRC staff members admitted that the number of new problems that had turned up at Zimmer would greatly delay the review of the quality of construction at the facility. Mr. Keppler, NRC regional director, told the commission that as much as 50 percent of the welding work was questionable and that there were still serious questions about the traceability of materials and the cable separation work. NRC chairman Nunzio Palladino was "highly distressed with the situation" at Zimmer and admitted that problems were surfacing "faster than the utility could handle them." After the meeting, commissioner Victor Gilinsky stated that it would "take years just to completely identify the problems" at the plant.

On November 12, 1982, citing a "widespread breakdown" in the plant's safety program, the NRC ordered an indefinite suspension of Zimmer. The utility's quality conformation program (QCP) revealed twenty-two "major construction deficiencies" that could have been prevented had there been a properly managed QA program. For the first time in the history of the nuclear industry, the NRC ordered a construction shutdown for safety issues in a nuclear facility over 90 percent complete. Before resumption of work, the commission noted the inadequacy of the management of the Zimmer project by CG&E and the utility's quality assurance program. Under the NRC order, the firm conducting the independent review would recommend steps to ensure conformance with the commission's regulations. Once the review was completed, CG&E had to report to Keppler on the course of action it intended to take. The regional director had to approve each component in the order before construction could resume.

Various intervenors were encouraged by the ruling. Thomas Devine, legal director of GAP, said that the NRC action was a major victory for "whistleblowers" who had brought Zimmer irregularities to light. Thomas Applegate was delighted but noted that opponents could not relax their vigilant efforts, while the attorney for Zimmer Area Citizens, Andrew Dennison, felt that the decision was appropriate, if not overdue. In general, most felt that strong measures were necessary to reassure the public, but some began to doubt whether the plant would ever go on line.

Instead of reassuring the public, CG&E's selection of an independent auditor produced unsettling questions. CG&E hired Bechtel Power Corporation, which had experience in nuclear power but had already been hired by CG&E to do a preliminary assessment of the Zimmer project. "The $64 question," said David Altman, "is how independent is the independent third party auditor going to be?"

The NRC demanded that Bechtel produce sworn statements indicating previous business deals with Zimmer or any financial interests in CG&E. On February 10, 1983, the NRC informed CG&E that Bechtel could either manage or review work, but it could not do both. CG&E retained Bechtel as a comanager of Zimmer and hired Torrey Pines Technology to oversee the plant's review.[52]

At this time, the Nuclear Regulatory Commission modified its July 30, 1982 decision refusing to reopen licensing hearings. The commission invited the MVPP to file its charges—an invitation that was viewed by some as a signal for new hearings. In June 1983, the MVPP reiterated its eight contentions raised earlier but criticized the NRC for its refusal to remove CG&E from control of the construction when its management problems were obvious. On July 12, 1980, the citizens' group filed three separate briefs with the NRC, its Atomic Safety and Licensing Board, and its Atomic Safety and Licensing Appeals Board. Listing fifty new allegations, the MVPP asked the NRC to reopen hearings on the nuclear power plant.

The MVPP's chances of success seemed enhanced by the board's earlier decision on May 2, which affirmed the licensing board's initial decision withholding authorization of a full-power operating license for Zimmer until CG&E had demonstrated reasonable assurance for adequate on-site/off-site emergency evacuation plans. The appeals board modified the requirement of a final report for FEMA but maintained that ZAC-ZACK and the city of Mentor were entitled to a later hearing on school evacuation plans.[53]

After a four-month investigation, Torrey Pines concluded that the power plant could probably be completed to conform with NRC regulations. However, the investigators revealed that CG&E was in complete disarray regarding quality assurance at Zimmer. The report was particularly critical of senior management and blamed them for a corporate policy that minimized expenditure and ignored quality and quality assurance. The review accused management of lacking involvement with and commitment to the project, as well as being ignorant of what was required to build a nuclear plant. Torrey Pines recommended the formation of an oversight committee; the hiring of an architect-engineer-contractor to replace CG&E construction management; the centralization of administrative activities at Zimmer; and the retention of another independent review organization to audit the quality assurance program.[54]

The Atomic Safety and Licensing Board decided not to reopen licensing hearings on safety issues at Zimmer. In a unanimous order, the board held that the MVPP had failed to file its contentions in a timely manner. Because all of the charges could have been made by 1981, the licensing board inferred that the MVPP's allegations merely refined earlier contentions.

Within days of that decision, the NRC removed James Cummings from his job as director of the agency's Office of Investigator and Auditor. Cummings, who had been actively involved in investigations of the Zimmer plant, had been criticized by United States District Judge Thomas Hogan for obstructing the release of responsive materials to GAP.

Shortly after the Cummings incident, the NRC scheduled a hearing in Cincinnati for November 1, 1983. At this time, CG&E was to present to the NRC its plan for managing and building the plant to conform with federal safety standards. Groups opposed to the Zimmer plant were free to submit testimony at a later meeting that same day. CG&E hoped that the NRC would approve its request to correct structural problems while finishing the construction of the power plant.

Once again the MVPP and GAP gathered evidence of newly discovered construction faults and prepared to ask the NRC to reject the utility's plan to finish Zimmer. Intervenors presented a list of fifty-six quality assurance problems at the plant. Again, the NRC ruled in favor of Cincinnati Gas and Electric. In spite of serious safety-related problems, the NRC agreed to allow the resumption of construction on the facility, which had been halted thirteen months earlier.

Reports began appearing in the news that CG&E was considering scrapping the project because an estimated additional $1.5 billion and two years of construction would still be needed to complete the project. The Ohio Public Utilities Commission also ordered a review of Zimmer in order to identify the costs that had resulted from mismanagement and those that had been justified. The utility was faced with spending $1.5 billion with no guarantee of recovering the money through rate increases, or with canceling and taking a write-off. CG&E officials were unwilling to make the final decision until Bechtel had completed its cost review.

On January 21, 1984, the president of CG&E announced that the company was halting construction. "There will be an immediate cessation of further construction expenditures on Zimmer nuclear facilities," said William Dickhoner. Instead, the plant would be converted to a high-pressure, steam-generating coal-fired power plant. What NRC delays and public protests would not achieve, the voices of financial institutions accomplished. Investment bankers and commercial banks were concerned about uncertain timetables and final costs. CG&E had also been struggling for several months with its partners, Dayton Power and Light and Columbus and Southern Ohio Electric, to negotiate a settlement of the financial losses caused by Zimmer. The concerns of potential financiers were "very important factors" in the willingness of CG&E and C&SO to end construction of the nuclear plant and to accept DP&L's proposal to convert to coal. CG&E indicated that the utility did have broad assurance that financial assistance was available if the industry initiated conversion plans. As long as the utility could demonstrate stability in its future planning, financial institutions could better calculate risk and lend money.

Critics of the plant were delighted. Dr. Fankhauser felt as if a great weight had been lifted from his shoulders. The MVPP agreed that the decision was the "wisest choice," and Thomas Devine of GAP reemphasized that it was a victory for whistleblowers. "Very simply," said Andrew Dennison of ZAC-ZACK, "from time to time the system does work in terms of citizens controlling their destiny." Although it seemed as if a major battle in the campaign against Zimmer was won, consumer advocates began marshaling forces to ensure that Ohio consumers would not bear the brunt of CG&E's mismanagement.

From the financial perspective of the intervenors, stopping Zimmer may well be a case of winning the battle and losing the war. As long as Zimmer is off line, Ohio ratepayers are spared cancellation costs. Once the conversion to coal takes place, it has been estimated that ratepayers will pay conversion costs, which may average twenty-three dollars on the monthly utility bill.

The financial impact of the coal conversion raises an interesting point about public intervention. Intervenors are not homogeneous. Opponents of Zimmer include antinuclear groups and consumer groups. Although the two groups may be aligned in their opposition to Zimmer as a nuclear facility, they diverge after that point. Antinuclear advocates will trade finances for safety, whereas consumer groups prefer lower costs. Public participation is not a monolithic force opposing utilities. This "one world" view is simply not the case. Rather, divergent interests exist that require accommodation.

REGULATORY STRUCTURE

The Zimmer case study demonstrates how regulator and regulatee are joined by public and private interest groups in the resolution of a complex public law dispute. The formal agency is the Ohio Public Utilities Commission (PUCO), created in 1961. The governor appoints five commissioners to the PUCO for five-year terms. These commissioners must have at least three years' experience in fields such as economics, law, engineering, or environmental studies. The general purpose of the PUCO is set out by statute, vesting the commission with the power and jurisdiction to supervise and regulate public utilities.[55] Ultimately, the commission's duties begin and end with protecting the public interest. However, because the commission is a creature of the state legislature, it cannot exercise any powers except those conferred on it by statute. The only court empowered to enjoin, restrain, or interfere with the commission is the Supreme Court of Ohio.[56] In fact, under the required "unlawful or unreasonable standard of review,"[57] the court will not reverse or modify an order of the PUCO where the record shows that the commission's determination is not manifestly against the weight of the evidence and is not so clearly unsupported by the record as to show misapprehension, mistake, or willful disregard of duty. Thus, as in the federal system, a great deal of power and discretion is given to the commission.

The PUCO has full jurisdiction to hear and determine all complaints a consumer may make against a utility regarding service. By law, such complaints must be in writing, and if they are based on reasonable grounds, the commission fixes a time for a hearing. Once a time has been set, the commission must notify complainants and the utility and publish notice in the newspaper.[58]

The PUCO must hold public hearings concerning all rate-increase applications. Upon application, the utility must publish the substance of the application in the newspaper. If there is no objection, the PUCO sets a hearing date. "Any person who may be adversely affected by the proceeding may intervene upon timely filing of a motion," and such intervention is then granted at the commission's discretion.[59] All parties and intervenors are granted ample rights of discovery.[60] Once a decision is made, an order is entered, which takes effect immediately. A rehearing can be applied for within thirty days of entry of the final order by any party who entered an appearance.[61] After timely filing of an application for a rehearing, the supreme court can review the order.

As utility rates have risen, more use has been made of these hearings and appeal procedures as utilities request more rate increases. During this plant cancellation period utilities have asked for higher rates of return on investment to attract new capital to pay for new plants under construction or planned. An increase in the rate of return means an increase in rates. In reaction to these escalating costs and to the unresponsiveness of the public utility commissions to the desires of the public for cost containment, grassroots consumer groups have been organized for the purpose of representing the consumer interest. The focus of these organizations is economic, and the goal is rate reform. In addition to grassroots groups, organizations have been created by statute to represent the public in ratemaking process.

In Ohio, the Office of the Consumers' Counsel (OCC) is such an organization. Created by Ohio law in 1977, the OCC is composed of attorneys whose primary function is that of a statutory representative of residential consumers.[62] "The consumers' counsel may appear before the public utilities commission as a representative of the residential consumers of any public utility," and "no person may be appointed consumers' counsel unless he is admitted to the practice of law in (Ohio)."[63] As a legal advocate for consumers, the OCC appears before the PUCO to take appropriate action concerning the quality of the costs of service. The jurisdiction of the OCC extends to every case involving the fixing of any rate charged for services by any public utility. The OCC intervenes only in PUCO cases. "At the request of one or more residential consumers, the OCC may represent those consumers whenever an application is made to the PUCO by any utility desiring to change any rate," or when a complaint has been filed that a rate exacted by a utility is unjust, unreasonable, or in violation of the law.[64] The OCC is funded by assessments made against the public utilities within the state.

Through the ratemaking process, the PUCO considers the costs attributable to providing service, and management policies and practices. The commission cannot allow inclusion in the rates of any expenses incurred by the utility through imprudent management policies or administrative practices. This issue has become the central focus of the OCC's involvement in nuclear plant cancellations, especially as it relates to excess capacity and whether more generating capacity is needed or wanted. According to the OCC, this agency is not antinuclear but rather anti-excessive costs for the ratepayer.[65]

Financing a long-term construction project presents a difficult trade-off between present and future consumers. This trade-off, sometimes referred to as the intergenerational equity issue, comes about because utilities must borrow money to build; they cannot build completely from retained earnings. Thus, the utility must either collect rates during construction or wait until the project is completed before charging customers. The longer the charge is delayed, the higher the ultimate rates become, because interest accumulates.[66]

The PUCO may, in its discretion, permit a reasonable allowance for construction work in progress (CWIP), but only if the project is at least 75 percent complete. CWIP includes the cost of borrowing money, as well as costs for physical materials and supplies. It benefits ratepayers to the extent that it lessens rate shock. CWIP works to the detriment of ratepayers, however, because they must pay a return on a plant that is not providing them service. It is the OCC's position that CWIP should not be included unless a utility can demonstrate that a severe economic hardship will occur because of a lack of real dollars.

The already complex funding problem is complicated when a plant is canceled. As noted, costs of canceled plants are not passed on to Ohio ratepayers.[67] This landmark decision was the result of OCC efforts to protect consumers. The PUCO was willing to grant Toledo Edison's request for a rate increase to cover the cost of system testing of its Davis-Besse nuclear plant. It was on this basis that the OCC intervened. The Ohio court determined that the unit was undergoing start-up testing with no beneficial service to ratepayers or any positive net electrical generation on a daily basis, and, thus, it was not used or useful in rendering service. "It is only proper that the investors' venture be found operational before they commence to recoup their capital outlays from the consumers."[68] The court reversed the PUCO order and denied inclusion of costs in the rate base.

The Ohio court went further with this theory in *Consumers' Counsel* v. *Public Utilities Commission*.[69] In this case, the Cleveland Electric Illuminating Company (CEI) joined a power pool, which started building four nuclear plants, but because of decreased demand and stringent standards after Three Mile Island, they terminated the facilities. The CEI applied for a rate increase in order to cover costs of the terminated nuclear plants. The PUCO recommended that the investment in the plants be amortized over a ten-year period. In its order the commission said that cancellation gives rise to a current cost, not a past loss, and rejection of amortization would have a very direct impact on the utility's financial performance. However, the court determined that the extraordinary loss could not be transformed into an ordinary operating expense and denied inclusion of costs in the rate base.

The OCC has played an important part in the most recent actions regarding Zimmer. It is the second-most-costly plant in the nation based on cost per kilowatt hour, costing $4,419/kwh as opposed to the industry average of $2,542/kwh. Events leading to the cancellation decision and to the involvement of the OCC can be traced back to 1979, when C&SO, the first partner in this venture, was to receive $51.6 million in CWIP costs. That began a pattern of requests when CG&E and Dayton Power and Light (DP&L), another partner, started collecting CWIP from consumers at a cost of approximately $185 million. These allowances were based on assurances that the plant was 75 percent complete—an allegation that has never been supported by fact.[70] The OCC requested a management audit of Zimmer costs in three pending rate cases in 1981. The purpose of the audit was to analyze costs attributable to mismanagement in order to keep those costs from being assessed to the ratepayers. According to the OCC, such costs could not be quantified without an audit. In 1982, the PUCO granted rate increases to CG&E and DP&L, allowing further recovery of CWIP. Further audit requests by the OCC followed, as did a request by Cincinnati's Environmental Advisory Council for public input into the choosing of the third-party auditor. A hearing was held on January 12, 1983, at the OCC's request, to exclude Zimmer costs from C&SO's rate base.

When Torrey Pines issued its report finding CG&E management at fault for safety-related problems and construction flaws, the OCC asked the PUCO to stop accrual of AFUDC on Zimmer and to deny its inclusion in the rate bases of the three utilities. This complaint was dismissed by the PUCO. The Torrey Pines report also triggered a resolution by the Cincinnati city council, which had removed itself from the Zimmer case. The city council resolved to join the OCC in opposition to any further rate cases asking ratepayers to pay for costs connected with CG&E's mismanagement of construction. In September of 1983, the OCC made a fifth request for a management audit, and one was finally granted in October.[71] It also requested to intervene in the audit, which was granted in a limited fashion, and moved to extend the time for its completion in order to allow for a more thorough investigation and analysis. This extension was denied. C&SO dropped Zimmer costs from its rate base, which opened the door for refunds to all of its customers. In January of 1984, the OCC formed a citizen's advisory committee, whose primary purpose was to study alternatives to completion of Zimmer. At the same time, the OCC filed a petition requesting the PUCO 1) to stop all Zimmer construction; 2) to stop accrual of costs associated with Zimmer; and 3) to require CG&E, DP&L and C&SO to provide studies, reports, and other information concerning the alternatives for Zimmer and to show cause why their recommended alternative was the least costly for consumers. The Cincinnati city council also passed a resolution at this time saying that Zimmer should not be completed as a nuclear power plant.

When the utilities announced their plans for conversion of Zimmer to a coal-burning plant, the OCC filed a motion asking the PUCO for an order to show cause for the companies to prove that conversion to coal was the least costly option to the ratepayers. Factors considered in determining costs to the ratepayer for conversion of Zimmer include construction costs, amortization, depreciation and federal taxes, operating costs, fuel costs, and replacement of parts at ten-year intervals. Increased costs of equity are also a cost to be considered in conversion, as are sunk costs for coal that accumulate during the conversion period.[72] At the same time, the OCC petitioned the commission to open separate accounting files to distinguish between nuclear construction and coal construction costs.[73]

In a complaint filed March 2, 1984,[74] the OCC outlined a number of issues that had been part of their past actions and listed several "new" items, emphasizing the fact that each month's delay increased the cost of Zimmer by approximately $15 million. In addition to allegations of negligence, misfeasance, nonfeasance, mismanagement, and imprudence the OCC requested that the commission 1) show cause why the utilities' alternative is the least costly best alternative for consumers; 2) order cessation of accrual costs associated with Zimmer; 3) order segregation of accounts; 4) order utilities to provide information regarding alternatives, including cost data, cost-benefit analysis, projected capacity, and available resources; 5) find that the actions of the companies regarding construction were and are unjust, unreasonable, and the result of negligence; 6) find that rates charged and proposed are or will be unjust and unreasonable; and 7) find that the practices of the utilities with respect to construction of Zimmer are or will be unjust or unreasonable.[75] According to OCC spokespersons, they also filed a Freedom of Information Act request to keep documents from being destroyed at Zimmer.[76]

According to consultants to the OCC's advisory committee, only 20 percent to 40 percent, or $340 million to $680 million, of the investment already sunk into Zimmer can be practically used under plans to convert the plant to coal. The companies' interest in conversion stems from a desire to recover some of these sunk costs that would otherwise be unrecoverable. Most of the actual building is useless, and a new coal-fired boiler would have to be built, because nuclear systems are generally overbuilt for coal-fired plants.

Charles Komanoff, a consultant to the OCC, says that building an entirely new coal plant would cost about the same as it would to convert, not even considering the $1.7 billion already spent, which suggests that little of what there is at Zimmer is really helpful in a coal-fired plant. Komanoff testified at an OCC hearing that depending on what the PUCO does, $1 million to $2.2 million worth of costs could be charged to the ratepayer, which would have a great bearing on the feasibility and desirability of conversion in the eyes of the consumer.[77]

The effect of cancellation and Ohio's rule against including cancellation costs in a utility's rate base is unsettling for a utility. Equally disturbing is the uncertainty surrounding how much of the sunk costs can be passed on to consumers after conversion. Increased uncertainty means increased risk, and increased risk means increased cost of money. To lower costs and allow utilities to borrow at comfortable rates, risk must be reduced. Further, the Zimmer principals needed to improve their financial health to be able to raise capital.[78]

Clearly, the pressure to convert is great. Ohio law refuses to allow ratepayers to pay for canceled nuclear projects. The Zimmer plant was canceled at a cost of $1.7 billion. Before the three utilities that own Zimmer can recover any of the $1.7 billion of sunk costs, they must complete or convert the plant to a coal-fired unit. After cancellation, the estimates for conversion were as high as an additional $1.8 billion. The utilities were also concerned that much of the $1.7 billion already spent would be disallowed because of management imprudency. With so much uncertainty surrounding the proposed conversion, the cost of capital to the utilities escalated, because with increased uncertainty comes increased risk.

In a successful effort to reduce uncertainty, the utilities, the Office of Consumers' Counsel, and several groups of consumers negotiated a settlement regarding the allocation of cancellation costs. The settlement did not resolve all issues; however, it did bring stability to the utilities. The settlement disallowed $861 million of the money already spent on Zimmer construction from being included in the rate base, and established a ceiling of $3.6 billion for conversion to coal. The PUCO accepted the settlement, and shortly after the utilities announced their conversion plans. The utilities were able to plan conversion because lenders made money available as uncertainty was reduced.

The Zimmer case study is a microcosm of the problems surrounding nuclear cancellations. Competing policies, divisions of opinion between official decision makers, an inadequate regulatory system, and the influence of the traditional model on transitional problems all contributed to uncertainty and delay in decision making. The Zimmer settlement is a unique resolution but an unsatisfactory solution to the cancellation costs-allocation problem. While it does have the advantage of bringing interested parties to a bargaining table, thus reducing transaction costs by bypassing administrative hearings, the settlement demonstrates the inadequacies of the old regime and the need for new forms of dispute resolution. The Zimmer settlement has serious inadequacies. Some participants, notably the city of Cincinnati, were dissatisfied with the settlement process and the terms of the settlement. The city believed that bargaining power was unfairly distributed in favor of the utilities. More important, the settlement is premised implicitly on the traditional model of ratemaking. The rate base method is still used to help utilities cover their costs. The PUCO has given no indication that rates will be set any differently, and the price of Zimmer electricity will still be pegged to the utilities' internal costs. There is no incentive mechanism for the utilities to keep costs below the $3.6 billion ceiling. Rather, the incentive is to have ratepayers absorb as much of the nuclear construction costs as possible.

SUMMATION

Although this discussion of the Zimmer cancellation cannot purport to be an exhaustive empirical study of public participation in utility proceedings, it is instructive. The stories behind Shoreham and Marble Hill have similar participants. Clearly, without public intervention by either proxy agencies or grassroots groups (sometimes by both), it is not inconceivable that these three plants would be on line as nuclear plants and would be generating electricity. It is now also true that if these plants were on line, Zimmer would be unsafe, Shoreham would lack an evacuation plan, and Marble Hill would add excess capacity. Public participation helped prevent these events from occurring.

Public participation appears in several guises and occurs in several fora. Some groups are public, some private, and some, such as the Cincinnati Advisory Council, are a hybrid. As in the case of Thomas Applegate, private individuals can have a significant impact when they are supported by more-organized groups. Also noteworthy is the nonmonolithic nature of public participation. Intervenors have different interests which conflict. The existence of conflict illuminates the need for a decision-making apparatus that allows the several voices to be heard and allows decisions to be made after competing interests and values are aired.

The roads for intervenors are many, not well traveled, and generally uphill. Intervenors are constrained by limited resources, limited participation, an antagonistic bureaucratic model of regulation, and an institutionalized mindset. They are hampered by procedural requirements of timeliness, burdens of proof, and a policy of judicial deference to administrative agencies. Further, opposition to intervenors is based on the not-unattractive argument that intervention is costly and therefore inefficient. Clearly, delay means costs. Equally clearly, in the case of Zimmer, for example, expedition means increased safety risks.

The inefficiency argument is premised on a narrow vision of bureaucracy and misses the point of democratic participation in public policy making and decision making. Agencies can be either economically efficient or politically responsive. They cannot be both all of the time.[79] The efficiency argument treats the bureaucracy as a rationalistic managerial entity. The inefficiency argument also concentrates on the short term. An alternative vision of bureaucracy treats agency decision making and policy making as participatory processes.

Public participation is the means through which citizens realize their citizenship,[80] give voice to their concerns,[81] and establish a dialogue about issues.[82] By its nature and by the constitutional structure of our polity, public participation is essential. That is true especially regarding issues that have external, transgenerational, and wide-ranging effects beyond the interests of regulator and regulatee. Nuclear power is scientifically and technologically complex and is not devoid of uncertainty. It is because of the existence of uncertainty in nuclear policy making that choices about nuclear power are fundamentally political choices that do not necessitate rationalistic, technocratic, managerial decision making on every nuclear issue. Rather, the political side of nuclear policy making requires broad participation as the means for reconciling competing values and claims.

PUCs tend to be dominated by regulated industries, although organized public participation is increasing.[83] The single greatest barrier to effective public participation is lack of financial resources. Costs such as fees for transcripts, reproduction of required material, and technical expertise hinder potential participants. Intervenor finances are in marked contrast to the greater resources of agencies and regulated industries in terms of personnel, expertise, funding, and overall organization. Delays, common in regulatory proceedings, are particularly burdensome to citizens' groups. Inadequate notice of pending proceedings and unrealistic time priorities further foreclose potential participation. In view of problems regarding technical experts and costs, a formal hearing is much less susceptible to widespread citizen participation.[84]

If effectiveness is gauged on the basis of numerical "results," intervenors have little influence on a utility's decision to cancel a plant. The five major reasons for cancellation, in their order of importance to the companies, include: 1) lower forecasted load growth; 2) financial constraints (an unfavorable market); 3) regulatory changes and uncertainty, which increase lead time and construction costs; 4) reversal of economic advantage—alternative generation sources become more favorable; and, 5) denial of certification.[85] However, if one considers effectiveness from a qualitative rather than an empirical view, then intervenors constitute an essential, if nonquantifiable, element in this process. The joint government-industry venture into nuclear power resulted in a multi-billion-dollar contribution of capital that makes backing out of nuclear difficult. Utility inexperience combined with a lack of proper governmental supervision resulted in innumerable construction problems and potential safety hazards. Intervenors kept these issues before both the public and regulatory agencies.

In the final analysis, financial considerations more than public participation influenced the decision to halt construction at Zimmer. Yet intervenors worked for ten years to focus NRC and public attention on serious safety and construction problems at the facility. Likewise, intervenors in Indiana, Kansas, Pennsylvania, and New York kept the cancellation-costs issue before decision makers. It may be impossible to verify empirically just how effective the efforts of intervenors are. John Woliver, counsel for one of the Zimmer intervenors, compared the entire series of events from 1975 to the cancellation decision to a bridge. Each part of the structure is necessary to complete the span; no part can be arbitrarily discarded. For him, the intervenors were essential segments of that bridge. Without them, the cancellation would never have occurred.[86]

The argument for public participation in bureaucratic decision making does not ineluctably follow from the structural design of our constitution. Instead, a mechanistic argument can be made that posits that each branch of government (together with the fourth branch of administrative agencies) has a specific job to do. Further, our government, and with it our society, will function better (more equitably) and more smoothly (more efficiently) if each branch does the job it is supposed to do. Therefore, the executive should enforce laws, the legislature should make laws, and the judiciary should decide legal disputes. It follows, therefore, that administrative agencies, as the expert delegates of Congress, should be the policy makers on specialized issues. Consequently, public participation is most proper at the polls, where it has an impact on the true lawmaking authority—state and federal legislators. Public participation before agencies, then, need not be particularly extensive, because the polls are the most democratic fora. Agencies should be given leeway and deference to fulfill their charge.

This view of government as a machine with each branch a discrete cog is too simplistic. The history of nuclear regulation should be enough to dispel this view. The mechanistic argument ignores legislative failure, regulatory failure, the lawmaking functions of each branch, the fluid and synergistic relationship of lawmaking and policy making, and, not least, the reality that more law is made in agencies that in the other branches combined. More than a nation of statutes,[87] we are a nation of regulations.

Instead of ascribing a mechanistic analogy to government and deducing a theory of limited public participation, public participation should be seen as the lifeblood of pluralistic democratic policy making in agencies as well as in Congress. The mechanistic analogy would nicely describe our government if the cogs worked smoothly. They do not. The executive makes law as much as, if not more than, it enforces it. The judiciary makes law in the process of deciding disputes. And, the legislature has given agencies so sweeping a mandate[88] as to raise questions about overly broad delegations of authority.[89] If public participation is confined to Congress and the judiciary or limited before agencies, then the public voice will be substantially missing from the lawmaking process. For government to work, participation must be pervasive, not isolated.

The nuclear case studies demonstrate the need for pervasive public participation. The promotional nuclear policy of the 1940s through the 1960s was the direct result of a tight alliance between government and industry. The joint venture was assisted by law and legal institutions at every level.

Even acknowledging the widespread public acceptance of the promotional policy throughout this period, the downside of institutionalizing policy must be acknowledged. The promotional policy was entrenched in every branch and in administrative agencies. When public acceptance withered, again attributing the withering to changed market and political conditions, the policy remained nonetheless. To change policy requires a major response at every level of government, and public participation plays an important and sustaining role in changing the entrenched policy.

4. Nuclear Power Market

INTRODUCTION

The unequivocal message from the marketplace is that the nuclear industry is dead.[1] Given the billions of dollars invested over the last four decades, one wonders if this obituary is premature. Nevertheless, the poor financial health of the industry is real. Although to some extent the electric industry as a whole has had a sickly financial period, nuclear utilities are particularly hard hit, because of the magnitude of their costs. The large amount of capital investment required to maintain construction of a nuclear project, particularly when cancellation is a real alternative, has put pressure on utilities to change the way they finance projects. In addition, the decline in economic enthusiasm for nuclear power coincides with an unfavorable political climate and must operate within legal and regulatory regimes based on market assumptions that no longer operate. Thus, unrest in financial, economic, political, and legal structures contributes to the industry's poor health.

This chapter concentrates on the market for nuclear power and explains the factual assumptions on which the market rests. The central insight about nuclear power and its regulation is the fact that the regulatory state created a market. The nuclear market did not and would not exist of its own accord. Further, the market that was created no longer functions smoothly. The consequence of nuclear market failure is an uneven and unfair distribution of risks and costs.

Because the nuclear industry is nonintegrated, that is, no single firm controls the entire fuel cycle from mining to electricity distribution, it is inaccurate to single out a single entity or group of entities for study. However, investor-owned utilities, which generate 78 percent of the country's electricity, are the primary actors in the nuclear financial drama.[2] Utilities order plants, then enter contracts with reactor vendors, architects and engineers, and contractors to equip, design, and build the plants. Consequently, without power plant orders, the bottom falls out of the nuclear industry. This bottoming out has happened. Therefore, to examine how a once pronuclear policy has been undermined, this chapter concentrates on the health of electric utilities in general, and nuclear utilities in particular.

Another clarification is necessary. There are no utilities exclusively committed to generating electricity from nuclear power. Instead, electric utilities wisely use a mix of fuels. Although utilities are not given a completely free hand in choosing fuels, they do have fuel choices available. The Powerplant and Industrial Fuel Use Act of 1978[3] prohibits new utilities from burning oil and natural gas and prohibits existing utilities from switching to those fuels. With oil and natural gas options removed, utilities can choose, within limits, to continue generating electricity from an old plant, build a new coal or new nuclear plant, or invest in conservation.[4] The health of nuclear utilities is thus interdependent with the health of the electric utility that has chosen to go nuclear, although utilities heavily invested in nuclear projects are not as financially healthy as nonnuclear utilities.[5]

Electricity generation has entered an entirely new era of rising energy production costs and increased competition. We enter this new era with caution because of the unknown, and we look back to our historical roots for guidance. Unfortunately, history, in this instance, has little to tell us, because the regulation of public utilities is based on an economic model that depends on an industry with declining costs. Today, the electric industry does not have the structural characteristics it once had. Instead, as this and the next chapter demonstrate, regulation is moving from a traditional capital expansion model to a postindustrial competitive model in response to market changes. As a prelude to the discussion of the economic, financial, and regulatory characteristics of nuclear-generated electricity, the nature of regulated industries will be briefly examined.

Regulated industries are caught between political and market forces. As regulated firms, electric utilities must serve two masters. They must satisfy the demands of the market in their attempts to raise capital, and they must satisfy the service obligation imposed upon them by government. The relationship between government and market is not discrete; both forces influence each other. Market actors (private investors) assess investment risk in part by gauging the regulatory climate. If regulators award utilities adequate return on capital investment, then utilities can attract money. Similarly, government regulators look to the market to assess such things as management prudency, financial risk, and rates of return. At other times, market forces and government obligations impose demands of their own on utilities, and these demands can and do conflict. The desire for high rates of return to reward shareholders, for example, conflicts with the desire of ratepayers for lower rates.

Thus, regulated utilities occupy an odd status in a capitalist democracy. Utilities are primarily privately owned and are publicly regulated. At the same time, they are measured against comparable competitive industries and are partially insulated from market risks. The private/market, public/nonmarket nature of nuclear utilities underscores the political and economic dimensions of utility regulation.

Government regulation and market competition find common ground in the ratemaking process. Ratemaking is the central connection between the state and a utility's financial health. If the ratemaking process is sound, the financial health of utilities will be sound, as well. This soundness claim does not mean that regulators must grant rate increases geared to the satisfaction of utilities. Rather, sound ratemaking is based on efficient market mimicking and equitable political judgements. Economic, political, and regulatory coordination is not currently the case. The contemporary regulatory scheme is based on nineteenth-century market assumptions as we approach the twenty-first century. Clearly, the ratemaking process must be modified dramatically. Otherwise, electric utilities and the future of nuclear power will continue to be subject to an anachronistic scheme of regulation.

Traditional ratemaking has induced utilities to overinvest in what have turned out to be unnecessary and expensive nuclear plants. The severe downturn in energy markets during the 1970s aggravated the weakening financial condition of utilities. As nuclear investments were found to be financially unattractive, plants were canceled. Canceled plants left billions of dollars of costs, and cost-allocation decisions were required to be made by state and federal regulators. Regrettably, government decision makers have no sound regulatory theory for handling unnecessary plant costs. Cancellation cost-allocation problems are confronting an antiquated regulatory system committed to capital expansion in a declining costs market. Today's electricity market is one of rising costs, slowing demand, and increasing competition. This confrontation between the old regime and a more competitive market has caused theorists to question the economic model and regulatory theory on which contemporary utility regulation is based. Such questioning of theoretical assumptions is endemic to a transitional period. In response to the industry's financial predicament, regulators have attempted to accommodate utilities with financial support. The accommodation response is based on historical ratemaking principles and can be seen as temporary. Traditional ratemaking has had a venerable one hundred-year history and must now move in a new direction during a period of maturation.

This chapter first describes the received economic theory behind the regulation of public utilities and then describes current deviations. Next, the financial realities of nuclear utilities are explored. Finally, ratemaking is explained. The chapter concludes by posing a dilemma. Traditional ratemaking has contributed to the mistaken investment in costly nuclear power. The dilemma is that if traditional cost-of-service ratemaking holds, that is, if ratepayers pay only for the electricity they receive, then utility shareholders who invested in government-protected utilities stand to lose much of their investment. On the other side of the equation, if shareholders are protected, then ratepayers pay and receive no electricity in return. Chapter five catalogues the federal and state responses to this dilemma between ratepayers and shareholders.

REGULATORY ECONOMICS

A situation in which numerous buyers and sellers exchange numerous substitutable goods so as to maximize productive and allocative efficiencies and encourage innovation, is the hallmark of a competitive economy.[6] Recent experience in the microprocessor industry demonstrates how the theory works in operation. The most desired products at the most desired prices stay in the hardware and software markets; other products drop out. However, not all markets are competitive, because some markets are imperfect.[7] Nuclear power has two glaring imperfections. First, as previously discussed, there would be no commercial nuclear power without government financial sponsorship. Government support is known as rationalization.[8] Although rationalization is not directly discussed, this book is about the long-term consequences of rationalization on the interaction of nuclear policy making and the legal institutions designed to promote it. Second, firms lack incentives to internalize harmful externalities, at least in the short run.[9] Firm *A*, according to theory, has little or no incentive to increase the costs of safety improvements over plant *B* if it means that higher costs reduce its profits. Nuclear power regulation is premised on correcting these two instances of market failure. Direct and indirect subsidies create and encourage entrants into the nuclear market, and safety regulations force a nuclear utility to internalize its externalities.

The regulation of electric utilities is based on another example of market failure. Historically, utilities were perceived as natural monopolies.[10] The economic sin of monopoly power is that a monopoly can reduce output, raise prices, and cause a loss of consumer surplus all at the same time.[11] In addition, legislatures believed that the production of electricity was in the public interest and that government regulation could assure its delivery at the lowest cost.[12] The importance of the public-interest rationale for utility regulation must be recognized. Utility regulation is at least as much political as it is economic.[13] It is impossible to talk about utility regulation in economic terms alone. Viewing electricity as a desirable commodity is a political justification for government regulation. The natural-monopoly rationale is the economic justification.

An electric utility is a highly capital-intensive venture. A utility must invest three or four dollars in a plant to produce one dollar of revenue.[14] In any given service area, a utility must build a generation station or buy electricity, and then must transmit the electricity through power lines to the ultimate users. If two utilities begin competing in the same service area, the loser will have invested unnecessary capital, because the additional plant and equipment will go unused, thus creating economic waste. One way to avoid duplication (waste) of fixed capital assets (plant and equipment) is to limit entry to a single provider. Economies of scale may permit only one optimum-size producer in a market; it then becomes highly desirable for public policy to allow a monopolistic supplier to operate.[15] In the case of utilities, entry was closed after a franchise area was given to a utility. Instead of promoting competition in the electric industry, the state simply conferred monopoly status on a firm, thus precluding new entrants. The effect of this strategy was to prevent short-term waste at the expense of long-term competition. Although it may seem counterintuitive, the harms attributed to natural monopolies were combated by state-protected monopolies.

In addition to the rationale regarding waste avoidance, monopoly status was conferred on utilities with the belief that a high-fixed-cost, capital-intensive industry enjoyed economies of scale. Consumers benefit, the theory goes, by encouraging firms to produce more (build more plants), and as they do, the cost per unit of output decreases. Another cost-reductive measure is technological improvement. As plant designs become more sophisticated and efficient, production costs decreased. Nuclear technology, for example, promised significant savings in the cost of fuel. Even though a nuclear plant is more costly to build than other plants, fuel savings greatly offset construction costs in the 1950s and 1960s; thus, nuclear utilities could enjoy economies of scale.

Therefore, several natural-monopoly characteristics in electricity (and in nuclear power) production can be identified: high fixed costs, long-run economies of scale, technological improvement, and a predictable pattern of growth in demand. Combined with an expanding economy and with a belief in a direct GNP-energy link, these characteristics had the effect of lowering the per unit price of a kilowatt of electricity. Simply, the larger the utility, the lower its production costs and the cheaper its product. Further, the more energy that was generated, the greater the Gross National Product. Large central power stations with monopoly protection performed well and seemed promising. This pattern was historically true into the mid-1960s.[16] These supply characteristics for electricity, together with the multiple oil price escalations of the 1970s and the country's desire for energy independence, made nuclear power attractive until well into the decade.

The market for electricity also has demand characteristics that have been used to justify regulation. The product of a nuclear utility, electricity, is treated by consumers as a necessary or highly desirable commodity in modern society. To the extent that consumers depend on this resource, they may find it difficult to substitute other resources as the price of electricity increases. Thus, consumers will pay more for less or the same amount of product. That means that electric prices are relatively inelastic, and price inelasticity of demand is another justification for regulation. An example of price inelasticity is that if the price of electricity increases by 100 percent, and the demand declines only by 10 percent, the relationship is said to be inelastic. The implications should be obvious. If the demand for electricity is inelastic as prices increase, then there is a greater transfer of wealth from consumers to producers, because some consumers have greater difficulty changing to alternative sources of electricity. The general wisdom is that the price elasticity of demand for electricity is relatively inelastic, with some signs of improving long-run elasticity.[17] Put another way, as prices rise, consumers will slowly put their money to other uses. Consumers will either use less electricity (conserve) or satisfy their electricity needs from other sources, such as investing in more energy-efficient appliances or acquiring electricity from alternative producers over the long term. Because of the relative inelasticity, there are cross-subsidization effects between classes of consumers. As prices rise, consumers who can move off the line will do so, leaving more costs to be spread among more captive customers. Managing cross-subsidization or distribution questions is also a justification for regulation.

Because consumers demand that electricity be readily available, and because electricity is virtually unstorable, there must be enough generating capacity available to meet demand. Utilities build two types of generating facilities—base-load and peak-load plants. Base-load plants (all nuclear plants are this type) are to be run continuously. Peak-load plants serve when demand surges, such as during the hottest day of the year. In order to meet their service obligation, utilities must be prepared to satisfy peak demand.

These demand characteristics, coupled with the industry's economies of scale and with a regulatory system that promotes capital expansion, mean that large plants will get built. Electricity supply and consumption worked together so well into the 1970s that the phrase "excess capacity" was rarely mentioned. Today the concept and its cause must be understood. Utilities have available for their customers a "reserve margin," which is an amount of electricity above estimated peak. "Excess capacity" is defined as the amount of capacity above the "reserve margin," which is roughly set at 20 percent of peak. Excess capacity has been calculated as high as 57 percent over peak.[18] When prices and costs of electricity began to turn upward and demand declined accordingly, the country found itself with more plants than it could use.

While some sections of the country have more electricity than others, the problems facing central power station-generated electricity remain the same. The cost for electricity being produced by central power stations is rising. As costs rise, competition increases. Competition in the electric industry comes in the form of conservation and alternative sources. Large industrial users can satisfy demand from cogeneration, self-generation, or bargaining for lower prices. Smaller, residential users can insulate or invest in solar power. In both cases, the alternatives compete with central power station electricity. That should put a downward pressure on prices. However, with cost-based ratemaking there is less incentive to compete. More important, captive customers will be forced to absorb more expenses.

One economist describes the electric industry's dramatic cost reversal as a normal part of any industry's life cycle.[19] Another study attributes cost increases to inefficient use of capital (underutilizing an existing plant); temporary exhaustion of economies of scale (no technical improvement); increased fuel costs; increased capital costs per kilowatt hour for new construction; and the occurrence of costs for plants later abandoned.[20] The undeniable conclusion is that the market for electricity is not what it was during its infancy and developmental periods.

This new, more costly market for electricity necessarily means more competition for investment dollars. Capital investment in electric utilities will become necessary as old plants are retired or refurbished and as new technological efficiencies are realized. The new market also means that nuclear mega-plants are no longer a given. Nuclear resurgence does not fit neatly into future energy projections.[21] Long lead times and large front-end construction costs for nuclear plants hit the upturn in electricity prices at precisely the most inopportune time for the industry. Not only are nuclear plants financially unattractive, all large-scale, costly plants are unattractive. Nuclear units will not be ordered until two economic conditions are met. First, the demand for electricity must require new large-scale plants. Second, nuclear plants must be cheaper to build and operate than coal-fired, oil, or gas-fueled units.[22] Still, with a market in which there is less reliance on central power station electricity, the role of nuclear power is circumscribed.

Historically, the desire for nuclear power fit neatly with electricity planning. Market, legal, and technological forces functioned to promote the development of this energy source. Something happened, however. The market, on both the supply and the demand sides, changed, and nuclear power was out of step with changing conditions. Contemporary supply-and-demand characteristics for the delivery of electricity have shaken the industry. No longer is demand growing predictably; no longer are economies of scale being experienced; and no longer are production costs declining. Further, the price elasticity of demand is showing some responsivity, which means that consumers will move off line when cheaper alternatives are available. The implication of these changing market characteristics is that the wisdom of a policy commitment to large central power stations, and a fortiori nuclear power, must be questioned.

As a matter of national energy planning, policy makers must decide whether large central power stations should continue as the predominant source of electricity. While projections indicate a current growth in the demand of just under 3 percent per annum, some industry watchers argue that new large plants will be needed by the mid-1990s.[23] Others suggest a variety of electricity production alternatives, ranging from small coal plants to mega-plants to nuclear resurgence.[24] The way to reconcile the existence and history of large central power stations and emerging competition is by abandoning allegiance to the traditional rate formula and by substituting a more competitive ratemaking scheme, which will be explored below. In this way, the mix of suppliers will be made by a regulatory scheme that is coordinated with contemporary market conditions.

FINANCIAL INDICATORS

Starting in the mid-1960s, with the rise of interest rates, production costs, and inflation, the electric industry began to feel financial palpitations. During the post-TMI period, utilities heavily invested in nuclear generation experienced severe financial strains, and specific projects have become financial fatalities. Currently, financial indicators show signs of recovery for those utilities not overinvested in nuclear power. Still, the palpitations and strains must be taken seriously. They have taken their toll on the industry and have altered the way utilities will be financed in the future.

The financial health of utilities is measured in the capital market. In June 1984 the United States General Accounting Office analyzed the financial health of the electric utility industry and gave it an overall positive prognosis with a cautionary warning about so-called nuclear utilities.[25] The GAO associated a utility's financial health with its ability to raise money, both debt and equity financing, and to earn a return acceptable to investors. Although the GAO identified seventeen financial indicators used by the industry, financial institutions, and PUCs, it chose three indicators for its analysis:

—rate of return on common equity;
—ratio of market price to book value of common stock; and
—corporate bond ratings.

Common stockholders invest in a utility with the belief that the investment will pay a return worth the risk. The utility investor expects to earn a return through dividends and appreciation comparable to that associated with other equity investments. The return on equity indicates a firm's success in attracting investors and measures how well a firm has done with the shareholders' dollars. Return on equity, however, will be paid only after other debt (including bond) obligations are paid. If revenues are insufficient to cover debt interest, dividends can be reduced or eliminated, and that has been the case with some nuclear utilities, such as Consumers Power and LILCO.

The ratio of market price to book value compares the market price of a firm's common stock with its book value, which is the amount of common equity divided by stock outstanding. This ratio is a gauge of a firm's future rate of return. A market-price-to-book-value ratio of one signals that the investment community assesses future returns to be strong, because the market price of the stock is equivalent with the firm's assets. A ratio of less than one indicates stock dilution (that is, the stock is worth less than its book value). Nuclear utility stocks have experienced dilution.[26]

Finally, corporate bond ratings measure a utility's ability to repay debt, whereas the other two measures assess equity. Bond ratings are developed by private firms to provide information to the financial community about the creditworthiness of individual companies. Because bonds are usually issued as long-term debt, the utility's rating acts as a future assessment of the company's strength.

The GAO report computes the return on equity and the market-price-to-book-value ratios from 1974 lows of 10.66 percent and .76 respectively to 1982 trends of 13.8 percent and .94. Bond ratings have demonstrated less resilience and have responded slowly to a general economic recovery. Still, economic signals are such that industry bond ratings should follow suit and improve.

Nuclear utilities have had a poor history of bond ratings during the past decade. This trend is crucial, because declining ratings discourage capital investors. A lower bond rating means higher costs of new debt capital and higher gross revenue demands. Even a short-term bond issue by a utility with less-than-perfect ratings can be a major strain on finances.[27] Public Service Company of New Hampshire, for example, borrowed $90 million of short-term financing in 1984 at 20 percent to help finance Seabrook. Ultimately, the difference in interest rates is calculated into the rate base and paid by the ratepayers.

The 1970s were hard years for the entire economy. High, rapid inflation and declining growth affected all industries. The electric industry, although certainly affected, was not among the most severely hit. Instead, it stayed near the middle and did not undergo the rapid fluctuations of other industries. The next chapter will describe how the industry's regulated status acted to alleviate what might otherwise have been a radical financial impact.

Utilities responding to current financial stresses have two principal options. One alternative is to stop construction until the financial environment is more conducive. This alternative is expensive. In 1981, for example, Union Electric Company canceled its $2.2 billion Calloway Unit No. 2. The resulting after-tax loss was $70 million.[28] After sinking so much money into construction projects, utilities find it extremely difficult to cancel and swallow costs when nothing is produced. Utilities then seek regulatory relief. The other option is to continue investing and convert to coal or finish the nuclear plant.

Nuclear utilities are less solid financially than electric utilities generally. Many nuclear utilities are carrying larger debts than they are used to, and those debts are greatly in excess of the types of debt carried by nonnuclear firms. Nuclear utilities also have a higher ratio of nonproductive to productive assets than other utilities. Further, utilities are not assured that PUCs will allow them to recover completely their debt costs through rates. This uncertainty allegedly has the effect of forcing managers to reduce or eliminate capital spending.[29] The reduction in prudent capital investment means that investors will look elsewhere for investment opportunities, thus taking capital away from utilities. In the next chapter, it will be argued that rate regulation still favors keeping electric utilities financially healthy and that PUCs are actively accommodating the competing interests of ratepayers and shareholders.

The primary difference between nuclear and nonnuclear utilities is the amount of money nuclear units have tied up in nonproductive capital assets. As long as a nuclear plant is under construction, it cannot produce electricity and cannot earn a full return on its investment. States that do not allow CWIP in the rate base give the utility no return until construction is complete. The costs of nuclear construction are such that they absorb a major portion of a firm's assets. Philadelphia Electric Company, for example, has assets of $8.1 billion, is capitalized at $7.57 billion, and has $3.4 billion tied up with the construction of Limerick Units 1 and 2. At first glance, it may seem odd that a utility would invest so much in a single project. After all, the more an investment portfolio is diversified, the greater stability there is against major fluctuations in a single investment. However, electric utilities, as regulated firms, operate in a constrained market. Through the ratemaking process they are encouraged, then rewarded, for investments in capital expansion. This incentive system works well in a market with declining costs, increasing demand, and promising technological developments. It functions poorly when the market is saturated, and costs start to rise because the industry has reached a turn in its S-curve.

Financial and capital forces have caused utilities to reorient their financial structure.

> Proper capital budgeting and financing decisions by both the utility and regulators are essential if the prescribed goal—adequate service at a minimum cost consistent with a reasonable return to investors—is to be realized.[30]

This quotation from a recent text on public utility economics and finances neatly captures the curious position of public utilities in our mixed market-regulatory economy. A firm in an unregulated market, to the extent such a market truly exists, is free to decide what product to produce and at what price to sell it. The unregulated firm takes the business risk that its product may not sell. Likewise, such a firm is free to decide how to finance its operations and how long to stay in business. Indeed, in competitive markets there is little relationship between the internal capital structure of a firm and product price when the firm is a price taker and the price is set in the market.

A firm can raise money internally from depreciation, retained earnings, or deferred taxes. Once those accounts are depleted or dangerously low, a firm can raise money from outside sources by issuing common or preferred stock or bonds or by short-term borrowing. The capital structure of a firm, the ratio of debt to equity, defines the financial risks the firm takes. In a competitive market, if the product the firm produces fails, the firm's owners (shareholders) lose their investment, because they assumed all business and financial risks, they chose the capital structure of the firm, and they chose the product and its price. This freedom of choosing a firm's capital structure is constrained only by what the market can bear. A firm cannot borrow at interest rates that raise the price of its product above competitive market levels, for example.

Utilities do not operate under the same constraints. Utilities earn money based on a rate of return calculated by averaging the amounts of debt and equity outstanding. A utility's capital structure directly influences product cost, because rates and the rate of return are based on the amount of the utility's outstanding debt and equity. As long as investment is prudent and the plant otherwise qualifies, financing costs are placed into the rate base. A prudently operated utility has little business risk, because demand is high (even inelastic) and the utility has a service obligation. Further, a utility has little financial risk because of cost-based ratemaking. There is a rub, however. The more rates increase in real terms, the more pressure is put on PUCs to examine the prudency of investment and to disallow questionable or costly investments. The invisible hand of the market is displaced by the political sensitivities of government officials, which thus increases a utility's financial risk.

The state has made the delivery of electricity a policy. As long as there is demand for the utility's product, the utility must satisfy that demand. This service obligation is an important constraint—the utility is required by law to sell its product. A utility cannot devote revenues to other investment opportunities until its service obligation is satisfied. Consumers are not bound by this constraint; they need not buy a utility's product. Consumer freedom, however, is circumscribed by an inelastic price elasticity of demand. Consumers are free to move, but utilities are not as free to diversify. That means that unless and until central power station electricity is the cheaper alternative, consumers will demand less. As the market for electricity grows more competitive, regulators must reevaluate the nature and extent of the service obligation.

Consequently, the utility, its owners, and its customers face a perplexing problem. As demand declines, prices rise, but the service obligation continues. The utility must stay in the electricity business, but consumers are free to shop around for alternatives subject to inelastic demand. Electricity prices are in an upward spiral. Demand (or the rate of increase of demand) declines, prices rise, the cost of money (financing) increases, prices rise, excess capacity increases, prices rise, "regulatory penalties" increase, prices rise, etc.[31] Breaking out of the increasing-costs spiral requires a rethinking of the traditional regulatory model. The traditional formula pegs a utility's rates to its costs. In a rising-costs, more competitive market, all financial risk is imposed on consumers under this formula. That is contrary to economic logic. As the market becomes more competitive, so should prices. Likewise, if financial risk continues to be passed through to ratepayers, inefficiency results, because waste is produced.

Thus, the relationship between financing and risk is not the same for public utilities as it is for competitive firms. First, a utility's financial structure plays a significant role in determining the level of rates. Second, most risk is allocated to ratepayers rather than shareholders. Third, regulators, not managers, set prices. Finally, the product of a regulated firm cannot fail as easily as the product of a private firm, because of the monopoly status and service obligation of the utility. For utilities, then, financial risk replaces business risk. Further, financial risk increases as utilities are forced to borrow more to cover interest obligations on long-term nuclear projects.

The financing of utilities is also experiencing the unevenness of the transition. Traditionally, the capital structure of utilities was highly leveraged; that is, a utility was financed with more debt than equity. Almost two-thirds of the money raised by utilities comes from bonds and preferred stock. High leveraging makes sense when the industry has expanding demand, declining costs, economies of scale, and a rate structure that rewards capital investment. Debt financing was encouraged. Utilities could borrow large sums and keep the firm's overall rate of return fairly low, because bonds are viewed usually as a less risky investment than common stock. High leveraging also has the effect of increasing the attractiveness of common stock by raising the return on common, because a smaller percentage of common stock is issued.[32]

Leveraging works less well in today's climate. Long lead times for construction projects, a period of double-digit inflation, and historically high interest charges mean that financing charges can (and have) approached the principal costs of construction projects. When these carrying charges are given rate-base treatment and allowed to earn a rate of return, consumers are paying interest on interest. High leveraging also means larger fixed costs and lower interest coverage, thus increasing the risk of bankruptcy.

Heavy carrying charges for utilities invested in nuclear plants have resulted in an overcommitment of capital to projects that do not produce revenue for some time. If a utility becomes too heavily leveraged, it will not earn enough revenue to pay its debt. The utility can (must) then try to raise money through equity financing. Moving financing from debt to equity during a delicate financial period means that the return on equity must be raised to attract capital, which again raises the cost of money. However, when internal sources of capital are exhausted and debt charges absorb income, the value of new common stock is diluted.

Both debt and equity financing have the effect of putting a cash-flow squeeze on utilities. The pressure for increased cash flow forces utilities to engage in creative accounting techniques through such devices as accelerated depreciation, normalization of taxes, and construction work in progress in the rate base. Thus, the financing of nuclear utilities is undergoing notable changes. Moving up in time the amount of depreciation a utility can take in a given year means that more money is available earlier for the utility. Likewise, tax normalization adds to cash flow by treating accelerated depreciation as a deferment of taxes. Finally, CWIP is an accounting entry that includes money invested in an ongoing construction project in the rate base, thus allowing a rate of return on money invested in construction even though no electricity is produced.

The express intent of each of these methods is to improve the cash-flow position of utilities. Cash-flow accounting was triggered by the changing market structure for electricity and was prompted by the inadequacies of traditional financing schemes and by traditional regulatory practices. These accounting methods are indicative of the change in the financial structure of utilities. If utilities cannot raise money to pay debt or attract investors, bankruptcy becomes a possibility. Because of the nature of electric utilities as regulated firms, bankruptcy is unlikely. Commercial lenders restructure debt because so much money is borrowed, and regulators grant emergency or temporary rate relief.

Although they are improving, the generally unfavorable financial indicators for nuclear utilities make the cost of money raising prohibitive in some instances. Further, cost increases have reversed the favorable economies of scale traditionally enjoyed by the industry. It is not inconceivable, or inherently undesirable, that inefficient utilities can or should go under and be replaced by other utilities or alternative technologies. In fact, electricity is being produced by other suppliers. Consistent with new suppliers, the financial community sees central power station utilities as a riskier investment than at any time other than the turbulent 1970s.

The substitution of unregulated producers for regulated utilities must be done cautiously if reliability problems are to be minimized. Further, the increase in deregulation must consider a quid pro quo for utilities. If their prices are to be set by more competitive standards, they must be given opportunities to diversify. These changes strongly suggest structural adjustments in a utility's corporate form, such as divestment by integrated firms of one or more segments of their fuel cycle. The promise of such changes is that the industry can achieve a higher level of competition (higher efficiency) than it currently enjoys. And, according to price equilibrium theory, the price of electricity will be set at its lowest cost. It is against this background that large capital-intensive central nuclear stations must compete.[33]

In conclusion, during the 1970s, utilities suffered declining earnings, lower bond ratings, market-to-book ratios of less than one, stock dilution, and an increase in the ratio of nonproductive to productive assets. These unfavorable financial indicators were aggravated for nuclear utilities. In the 1980s, these trends began to reverse as plants were completed or taken off line and, more significantly, as managers and regulators responded to the cash-flow needs of utilities. However, the transition period has noted its presence by altering the way capital markets and managers assess the financial structure of utilities. For the forseeable future, utilities must accept the reality of greater financial risk. Likewise, regulators must reassess the role of financial risk in the ratemaking process.

RATEMAKING

The political response to the problems of natural monopolies—lower output, higher prices, lost consumer surplus, and economic waste—was government regulation. The delicate problem was one of design. How can government regulation appear attractive enough to encourage the private sector to invest money in a regulated industry? The answer was to design a regulatory scheme that gave the utility some of the economic benefits of a monopoly without imposing a monopoly's costs on society.

Public utility regulation was based on a fundamental trade-off. In exchange for exclusive jurisdiction over a specific geographic service area, a regulated utility had to do two things. First, the utility undertook a service obligation, which meant that the utility could not move its resources into more financially attractive investment opportunities whenever they came along. Instead, investment had to be made in satisfaction of the service obligation. But what was to prevent the utility from shutting down service and going out of business? Surely, a "service obligation" cannot force someone to work without a financial incentive. That is true to a limited extent, and this need for "financial integrity" implicates the second part of the trade. The second, and more important, obligation assumed by a utility was to let the state set its prices. This price-setting function is also known as ratemaking. Alone, the idea of ratemaking by government rather than by market appears to be a disincentive. That is true until the ratemaking formula is examined.

The classic ratemaking formula can be stated as $R = O + B(r)$.[34] The variables are:

R = revenue requirement
O = operating expenses
B = rate base
r = rate of return

The formula was intended to duplicate certain market characteristics, and the formula worked exceedingly well for nearly one hundred years.[35] The revenue requirement *(R)* is the total amount of money a utility is entitled to earn by law. In other words, *R* is the amount of money a utility can charge its customers. In order to make a profit, any firm must recover its costs. Therefore, operating expenses *(O)* are recovered as long as these expenses have been reasonably incurred. In addition to operating expenses, which basically cover the utility's variable costs, the utility is entitled to recover its fixed capital investments. This variable is known as the rate base *(B)*. Not only does the utility recover its capital investment, it also earns a return on the investment *(r)*. These two variables *B* and *r* drive the ratemaking formula, because these are the variables that make a regulated public utility most like a nonregulated competitive firm. A utility manager can earn a profit for the firm by realizing efficiency in production. Further, capital investment, perhaps overinvestment, is encouraged, because the return on investment increases with the amount invested. The tendency of the rate formula to encourage overinvestment is known as the Averch-Johnson (A-J) effect.[36] The A-J effect appears to occur in the kind of declining-costs market described earlier. It most likely will not occur, or will be reversed, in a rising-costs market with the traditional formula.[37]

With some important refinements, public utilities were seen as particularly safe investments precisely because of their regulated status. The state gives a utility a monopoly *and* requires the utility to serve. In exchange, the state regulates rates according to a formula that encourages capital investment. Crudely put, the more money a utility spends, particularly the more money it puts into the rate base, the more money the utility earns. Although the state does not "guarantee" that a utility will earn a reasonable rate of return, as long as a utility is prudently run it has the "opportunity" to make money.[38] To a degree, that opportunity is protected by law under the constitutional prohibition against taking private property for public use without just compensation. An important qualification on this constitutional standard requires a balancing of shareholder and ratepayer interests. Today, that balance is being worked out. Shareholders want their investment protected. Ratepayers want reasonably priced service. If utilities cannot provide the service cheaply and reliably, they will not be rewarded.

Utility ratemaking calculations involve four basic steps. Initially, gross utility revenues are taken from company books. Because the revenue requirement is derived from accounting data, accounting entries such as AFUDC, CWIP, and depreciation are significant. Next, operating expenses, funds spent in order to receive gross revenues, are determined, also using figures from company books. The third calculation determines the rate base, an assessment of utility property actually providing service. The evaluation of on-line facilities determines what rates are charged and on what facilities a return should be made. The rate base generally includes only those properties that are "used and useful" in electricity production.[39] Finally, the appropriate rate of return is determined by creating a percentage multiplier of the rate base that produces the capital return to which investors are entitled.[40]

Gross revenues and operating expenses are the least complicated components of the ratemaking process. While the gross-revenue figure is derived directly from utility accounts, occasional debate surrounds what line items should or should not be included in the operating-expense category. Operating expenses are paid by utility customers directly. Fixed overhead, annual depreciation, and taxes constitute total operating expenses. Costs pertaining to production, transmission, and distribution, expenses resulting from fuel and labor costs, material and supply costs, and maintenance costs are also recoverable expenses.

Taxes are included with operating expenses because they are regular expenditures for a utility. The tax rate calculated by utilities often does not match the tax rate paid to the government. Utilities charge ratepayers a tax fee of about 48 percent. Yet, utilities actually pay a significantly smaller amount. In 1975, the rate of taxation was below 2 percent.[41] While this system contributes much-needed capital to utility reserves, the process does not reflect an even method for raising liquid assets. The accounting method of tax normalization is a way to defer taxes and increase a utility's cash account.

Depreciation is passed through to consumers as operating expenses. Plant property is generally between 80 percent and 90 percent of a utility's total assets; large portions of utility capital are invested in fixed, immovable property such as land, buildings, large structures, and machinery.[42] In order to maximize their financial position, utilities tend to put cash in property acquisition and facility construction and to depreciate the investment. Utilities earn a return on the undepreciated value of investment.

The uniform system of accounts defines depreciation as "the loss in service value not restored by current maintenance . . . and against which the utility is not protected by insurance."[43] Wear and tear, decay, action of the elements, inadequacy, obsolescence, changes in the state of the art, changes in demand, and changes in regulation are used to calculate depreciation expenses, which are then amortized over the life of the plant. Deterioration of the physical plant is considered simply a cost of doing business.

Depreciation has different financial impacts depending on the method by which it is recovered. Through straight-line depreciation, capital investment minus the estimated salvage value is allocated in equal amounts over the useful life of the plant. With straight-line depreciation, less cash is available for utility use throughout the life of the plant, because ratepayers contribute less in each particular installment. When pressed for cash, accelerated depreciation affords larger front-end cash payments, thus contributing to a utility's cash flow.

Little debate surrounds the issue of utilities' regaining interest expenditures from the ratepayers. The crucial issue is the timing of that recovery. Utility management argues that the immediate recovery of the carrying costs for plant construction is necessary to keeping utility finances healthy. Longer construction periods and larger construction budgets make CWIP a significant consideration for utilities. By 1974, CWIP was 25 percent of total plant construction costs.[44] To cope with heavy financing requirements, a utility's ideal scheme inserts all CWIP directly into the rate base. This immediate cash input improves cash flow. Utilities assert that ready internal cash is necessary and that severe financial strain, inability to finance capital needs, and inability to maintain adequate return result when CWIP is not included in the rate base. A further argument for CWIP in the rate base is that "true" price signals are given to consumers. This argument is more rhetorical than economic. By including CWIP in the rate base, present customers are subsidizing future customers. This intergenerational equity issue is countered by utilities who say that present customers are benefiting from reliable service and that CWIP reduces future rate shock.

Opponents of CWIP argue that there is no reason for ratepayers to contribute money to a project that may never be completed. Having no quibble with paying carrying costs when facilities are finished, opponents generally support an alternative method, Allowance for Funds Used During Construction (AFUDC). The carrying costs and all other expenses incurred during construction are recorded monthly. When the plant goes on line, these recorded costs are put in the rate base and depreciated over the life of the plant. Under the AFUDC model, payment of construction expenses begins once the plant starts operating. If a plant is canceled before completion, under the AFUDC approach, utilities will receive no return on their investment. Both construction and carrying costs will be lost. AFUDC thus restores utility expenses only after a plant is working, and construction capital is regained by the utility only after the project is completed.

Not paying for nuclear plants until completion may be appealing considering the current rash of plant cancellations. The harsh reality is that deferred carrying costs incur interest and may produce rate shock when a project is completed. Even if rates are phased in over a period of years, interest charges continue to mount; consumers pay interest on the principal investment, interest during construction when money is borrowed to pay debts, and interest during the phase-in period.

The justification of CWIP or AFUDC accounting procedures is more a matter of policy than of economics. There is little disagreement about the economic effects of the two methods. Rather, the disagreement is about the fairness of one approach or the other.

PUCs also assess the rate of return, which is the amount a utility is allowed to earn beyond operating expenses, depreciation, and taxes. Fluctuations in the rate of return can result in significantly higher or lower total earnings for the utility than even a large change in rate base. "A public utility is entitled to such rates as will permit it to earn a return on the value of the property which it employs for the convenience of the public."[45] The rate of return determines the amount of capital moving into the utility and performs a capital-attraction function. The return must be sufficient to maintain "confidence in the financial integrity of the enterprise, so as to maintain its credit and to attract capital."[46] The rate of return must be high enough to attract individuals who will contribute capital to utility reserves by purchasing bonds or stocks. At the same time, the return must be low enough to provide reasonable rates for consumers. Return reflects the cost of capital, the actual payments to past and present security holders, and can be used to measure the average cost of new capital. Determining rate base and rate of return is critical for the cash flow of utilities under the traditional method.

The cash squeeze on nuclear utilities is acute because of increased interest rates, increased construction costs, construction delays, and design changes.[47] Through the cash-flow accounting methods already discussed, utilities can receive revenue during plant construction and operation. If a utility has inadequate internal cash and the cost of money is too high, utilities look to the ratemaking process for financial help during construction. If a plant is canceled, utilities have asked for PUC relief through ratemaking.

The primary procedure is amortization of capital investment for failed projects. Amortization means that utilities may recover a project's costs in installments over a given period of time. If full amortization is allowed, at the end of the amortization period every cent of the initial costs plus interest is repaid to the utility. Most regulatory commissions have allowed utilities with canceled plants to regain a significant portion of their expenses through amortization. In most instances no rate of return is allowed on the unamortized balance, which forces investors to suffer carrying costs.[48]

Amortization is an attempt to include as operating expenses costs otherwise included in the rate base. Spreading the costs over a longer time period lessens the impact on the ratepayer.[49] Amortization results in ready cash. The rationale for amortization is in the recognition that at each stage of a utility's decision-making process, choices were prudently made. When utilities serve electric customers in good faith, the utility is allowed to recover initial investment costs and, perhaps, capital costs.[50] The use of amortization is a compromise position in reaction to the strict application of the "used and useful" approach, allowing no recovery when there is no functioning plant.

Tax treatment of canceled projects also affects cash availability. When a power station is canceled, utilities write off their sunk investments as tax losses for that year. The decision to cancel thus results in an immediate tax saving of significant proportions. The credits represent a burden to the taxpayer of up to 40 percent of abandonment costs.[51] With flow-through accounting, these tax breaks are transferred to ratepayers. Utilities using tax normalization, however, defer taxes and keep the realized cash.

Nuclear utilities confronted the world of higher costs and less demand at a time when they least needed it. The government-industry partnership that actively encouraged the growth and development of nuclear power was aided and abetted by a regulatory regime that contained a ratemaking formula committed to capital growth and development. As the demand and supply characteristics of the electricity market changed, regulators began to question the wisdom of the traditional rate formula in response to the voice of ratepayers and economists critical of the former rubber-stamp attitude exhibited by PUCs. The rate formula is not, nor was it ever intended to be, an automatic pass-through mechanism. Not all utility investment was to be financed through rates; otherwise utility investment is risk-free. Rather, a utility is rewarded only for "prudently" made investments that are "used and useful." Consumers were to be charged only for those investments from which they realized some benefit. Theoretically, if a plant is never completed, then the ratepayers are not to be charged for it.

Nuclear plant cancellations, and other large-scale cancellations, have greatly challenged that theory. Because of their size, and because of the length of time it takes to put a nuclear plant on line, vast sums of money are expended before any electric service is realized. PUCs are thus faced with the question of whether to follow traditional cost-of-service ratemaking and see a utility go under when a multi-billion-dollar investment is scrapped, or to help salvage a utility that has overinvested in a nuclear plant. In chapter five, the various federal and state responses to this problem are detailed. Suffice it to say that traditional ratemaking has outlived its usefulness. The creation of monopolies and ratemaking based on a model of capital expansion worked as long as costs declined. However, once costs began to rise, the traditional model responded slowly and sluggishly, leaving nuclear utilities in a financially precarious position. Ratemaking need not abandon its market-mimicking foundation. Rates can be based on the more competitive market that is developing rather than serve as a reward system for a utility's capital expenditures. Public utility regulators and regulations must confront the new industry dimensions just as financial markets must. Traditional regulation is anachronistic as cheap energy and declining costs are replaced with more expensive and costly energy sources.

Critics of PUC regulation argue that the current system produces what have been termed "regulatory penalties."[52] Critics argue that the ban on burning oil or natural gas in new electric plants necessarily places a cost on consumers if those fuels are cheaper than coal or uranium. The higher cost of fuel, then, is seen as a regulatory penalty. The second so-called regulatory penalty is the cost-of-capital penalty. That means that PUCs raise the cost of money in the marketplace when they refuse to allow firms to pass all costs on to ratepayers and instead force losses on shareholders. The direct consequence of this strategy is to force up the costs of capital, because utility investment is more risky, and riskier investments require higher returns.

Finally, consumers are threatened by a reliability penalty to the extent that utilities now underinvest in plants, because of the fear that PUCs will not give them their due.

Essentially, this fear of regulatory penalties arises from either a lack of confidence in PUC competence or a fundamental disagreement about the role of the PUC. PUCs not only set rates that represent a fair return on a firm's investment, but they also confront equity issues, such as which classes of consumers pay which portion of the rates. Rate design clearly has wealth redistribution effects, and PUC regulations concern themselves with those effects. The "regulatory penalties" are penalties only to those who limit the function of ratemaking to capital attraction, and who advocate the preeminence of electricity generated by large central power stations. PUCs have shown such favoritism in the past but need do so no longer as alternatives develop.

Assuming that regulatory failure does exist, two responses are available. First, deregulation, by definition, eliminates administrative costs and may seem more efficient. However, deregulation necessarily forces large classes of consumers at the retail level to absorb a utility's costs because they are unable to switch suppliers or conserve. There is little evidence that the electric market is competitive at the retail level for small consumers. Unless competition exists, deregulation will have drastic distributional consequences. Second, instead of deregulation, regulatory reform can take place. The key regulatory reform is to revamp the rate formula so that it is coordinated with the modern market. Ratemaking must abandon the traditional capital-expansion model and replace it with a competitive efficiency model. Neither industry nor ratepayers can rely on cost-of-service ratemaking, because it is too expensive. Cost-of-service ratemaking insulates shareholders from business and financial risks and concomitantly burdens ratepayers.

The most significant contemporary market change is the increase in competition for some segments of the industry as a result of rising costs. As the market becomes more competitive, it is mandatory, for efficiency and equity reasons, for rate setting to reflect the economic (market) value of a kilowatt rather than a utility's internal historical costs. By setting rates on a utility's avoided costs, that is, the cost the utility would pay to purchase a unit of electricity in the market, efficiency and equity in electricity pricing are realized. First, business and financial risks are placed on the utility. If a new project costs more than the economic value of electricity, then shareholders must suffer. However, the incentive behind the avoided-cost method is to encourage the production of electricity below market and thereby reward shareholders for efficiency gains. Second, intergenerational equity problems are avoided, because the price is set at current economic value. Finally, a ratemaking formula based on avoided costs privatizes risks and gains without socializing any single utility's internal inefficiencies.[53]

This market model for electricity pricing will not develop overnight. Until small users achieve purchasing independence, the need for a power grid and the need for large power stations will continue. As long as electricity is a desirable commodity, and as long as captive customers remain, retail distribution of electricity presents regulatory and reliability problems.

Increased competition in the industry is occurring, although not evenly across all segments. Increased competition will also affect the corporate form for utilities. Most utilities are integrated; that is, they generate, transmit, and distribute electricity. Competition at the generation point does not mean competition for retail sales. A change in ratemaking to a more market-based formula can result in utilities' undergoing corporate reorganizations that divest some segments for competitive purposes and keep others for regulatory purposes. During the industry's maturing period, firms will struggle to maintain their market shares by determining their proper competitive/regulatory mix.

This movement from regulation to deregulation brings the industry full circle. At this point, it is unlikely that regulators will step aside and let the market set all electricity prices. Nevertheless, the threat and promise of more competition is affecting the regulation of the industry. PUCs are beginning to play with redesigning the rate formula to incorporate the incentives of a competitive market. Rate formulas that avoid the Averch-Johnson effect and that promote energy or economic efficiency are being developed. Utility performance standards are also being developed to encourage firms to spend money prudently instead of automatically dumping capital into expansion. In the face of workable competition, regulation is wasteful. For nuclear power, deregulation will not be seen for radiological matters. Still, electricity pricing can undergo some deregulation.[54]

In contemplation of deregulation, PUCs are engaging in transitional ratemaking schemes that move away from rate base methodologies and toward market pricing. Traditional ratemaking methodology exhibits a lack of control and uncertainty in long-range utility financing. Rate reform proposals are moving away from rate base methods and are pegging prices to external standards designed to promote efficiency and equity.

CONCLUSION

The economic, financial, and regulatory status of electric utilities is in a period of great fluctuation. Because the structure of the market has changed, both the way that utilities structure themselves financially and how PUCs set rates are being altered to cope with market changes. The shift in economic, financial, and regulatory structures is profound. Electric utilities, particularly nuclear utilities, are forced to realize that the traditional capital-expansion view of the world can no longer be taken for granted. Instead, utilities are losing their preeminent position as suppliers of electricity.

Utilities must respond to the changing demand characteristics of a postindustrial economy.

The movement from the traditional model of electric utility regulation to a postindustrial model is manifest in a turbulent transition. The natural-monopoly concept is challenged by competition,[55] thus forcing policy makers to question the basic and historical economic model of utility regulation. Next, the use of highly leveraged financing has put nuclear utilities in a severe credit crunch, forcing them to engage in short-term accounting techniques expressly for the purpose of stimulating cash flow. This desire for increased cash flow has even caused some utilities to contemplate bankruptcy as a way of improving their cash position. Finally, the impact of changes in economics and finances is being clearly felt in the ratemaking process. The traditional formula, based on a historical economic model and influenced by a traditional financial structure, is no longer adequate.

The place of nuclear utilities in this changing market is uncertain. One thing is clear: Nuclear utilities will be forced to compete and will not enjoy the regulatory subsidies they once did. A scenario in which nuclear power plays a part in our electricity future is conceivable. In fact, some of the signs of resuscitation are present. If we continue to experience growth in demand, other fuel sources become more costly, public uneasiness about nuclear safety quells, and utility management becomes more proficient in building, then nuclear plants will be built. How likely is this scenario? In other words, nuclear power will not reemerge until economic and political indicators are aligned with contemporary conditions.

Growth in electricity demand can be taken as a given. However, regulated electric utilities can be less sanguine about their role in supplying that product. Conservation and energy efficiency are becoming increasingly attractive. Unless utilities invest in conservation or increase energy efficiency, their role as suppliers will be diminished, and utilities will have less of a market for their product. Currently, utilities are prohibited from using oil and natural gas as a fuel to generate electricity. Hydro-power capacity is virtually exhausted. That leaves only coal and nuclear. Today, it is cheaper to build a coal plant than a nuclear plant, and coal is in relative abundance. The prudent managerial decision is to go coal rather than nuclear. But things change. Coal can (and will) become more expensive as sensitivity to the problems that fall with acid rain and rise with carbon dioxide build-up become more acute. Public perception of risk naturally affects policy. If acid rain is perceived as threatening, then regulatory measures will be promulgated to force utilities to internalize these costs. Simultaneously, the public perception of nuclear safety is undergoing a period of quiescence. The restart of TMI-1[56] after eight years and the low-power testing of Shoreham[57] signal the NRC's willingness to proceed with nuclear power. The public voice has no effect until official decision makers give it effect. Public opposition has not stalled these generating stations from moving toward on-line status.

Finally, utilities, if they are to invest in nuclear stations, must learn to build them more cheaply. That can occur through standardization[58] or tighter managerial controls, or both, as cost-reduction measures.[59]

Nuclear utilities did not create this new environment. Rather, they are the most visible and logical consequences of a capitalist and industrial expansion ideology. Both the economic model and the regulatory response to that model are artificial constructs that are based on a policy commitment to large, high-technology, central power stations. Clearly, nuclear power plants fit the image. Equally clearly, that image is wrong in a more competitive market, and wrong in a political milieu not uniformly committed to nuclear power. Utilities can begin to cope with unanticipated expenses in a thoughtful, planned manner, and nuclear utilities will be given the opportunity to reemerge in a newly evolving, more competitive postindustrial market. The remaining question is, What should be done with the massive, mistaken investments in nuclear power? We turn to that question next.

5. Accommodating Nuclear Power

INTRODUCTION

The economic and financial situation of nuclear utilities as regulated industries was explored in the previous chapter. Clearly, utilities are caught between the free market and government control. As such, they may be in an impossible situation. Their regulation is not solely motivated by economic principles. Nor is it motivated by overtly political preferences. Rather, utility regulation is charged with balancing the economic and political interests of utilities against the interests of their consumers.[1] Utility regulation thus comprises both private/economic and public/political dimensions. In an volatile political economy, such as that being experienced by the nuclear industry, decisions and policies regarding nuclear power will lack uniformity as a new regulatory theory takes shape. The current condition of nuclear regulation, as a period in search of a theory, will be demonstrated by discussing the lack of uniformity among nuclear regulators in their cost-allocation decision making.

The cost-allocation problem for canceled plants is the most serious problem facing the nuclear industry, and it has a significant impact on the electric industry and on regulated industries more generally. Although regulatory authorities have faced cancellations before with other sorts of projects, [2] what differentiates nuclear plant cancellations is their magnitude.[3] Long lead times and massive investments require significant capital contributions over the construction life of a project. The failure to recoup these investments threatens to bankrupt utilities that have large commitments in nuclear projects. Even though bankruptcy is unlikely, rate shock is not. The need for capital put pressure on utilities to improve their cash flow and resulted in the accounting techniques mentioned in the last chapter. Utilities, in turn, have put pressure on regulators for financial help. For the most part, they have received that help.

Curiously, for a regulatory system that has served industry and consumers well for almost a century, regulators and existing regulations are poorly equipped to deal with failed nuclear plants. Although regulations exist for cost allocation, and although cost-allocation decisions are being made, these regulations are inadequate, and the decisions lack uniformity. These two infirmities of the regulatory process, inadequate regulations and lack of uniformity in decisions, share a common attribute—there is no solid regulatory theory for allocating the cost of large cancellations that threaten to bankrupt a utility.

This lack of theory has occurred precisely because the market for the delivery of electricity is changing dramatically. The market assumptions and the economic models derived from those assumptions upon which the electric (and nuclear) regulatory regime has been based, no longer obtain.[4] Current market changes are profound enough to force us to challenge regulatory assumptions seriously. At the simplest level, the question confronting the regulatory system is whether regulation should continue to encourage and support large-scale central power stations, including nuclear stations. The alternative is to let the electric (and nuclear) industry move into a more competitive market and out from under the state's protective monopoly mantle. The electric industry is experiencing a new stage in its development, a more maturing stage. The development of large-scale electric power coincided neatly with large-scale industrial development. As the country shifts away from an industrial economy, the supply of electricity must shift similarly. Changes in the market mean that utilities and PUCs must adapt. The delivery of electricity must now occur in a postindustrial world.

Nuclear power plant cancellations did not precipitate, but do highlight, these market changes. The cancellations have come about because of changes in the market. The development and promotion of nuclear power were consistent with market signals and, more important, were consistent with and encouraged by the legal system. Unfortunately, with nuclear policy, market changes precede changes in the legal system, which means that existing legal rules are applied awkwardly.

This chapter is about how the legal system deals with the problems of mistaken nuclear investment within its framework of existing rules. Specifically, the chapter examines the state and federal responses to nuclear plant cancellations and some attendant problems. Although the regulatory system's response has identifiable attributes, that response is inadequate. This chapter will first briefly identify the general attributes of the regulatory response to nuclear plant cancellations. Second, the discussion of state and federal cases, statutes, and regulations will demonstrate how particular decisions conform to the general attributes. Finally, the chapter concludes by evaluating the adequacy of this response and finds it wanting.

THE DUALISTIC RESPONSE TO PLANT CANCELLATIONS

The theme of this book is that nuclear regulation is undergoing a radical transformation. The persistent evidence of that transformation is that nuclear power and its regulation are stuck within several conflicting dualities. Previous chapters have explored the duality between centralization and decentralization and the duality between safety and finances. In the last chapter, capital-expansion ratemaking confronted competitive ratemaking. In this chapter, these three dualities are joined by another; the desire to protect consumers (ratepayers) confronts the perceived need to shelter investors (shareholders). Although this duality may appear easy to balance with a "split the difference" attitude institutionalized as policy, it is impossible to reconcile as a matter of economic theory given changing market structures. No single economic justification for accommodating the interests of shareholders and ratepayers exists. The assertion that no economic theory exists is based on the realization that the claims of shareholders and ratepayers are, at bottom, mutually exclusive.

There is no regulatory economic model, no matter how loosely one uses the term *economic*, that can continually and consistently reconcile the competing claims of shareholders and ratepayers. The one reconciliation that does work is essentially political and dispenses with the necessity of developing a theory of regulation dependent upon an economic calculus. This political response will be presented here as "accommodationist." The accommodationist response is temporary and transitional. Its intended purpose is to allocate losses between ratepayers and shareholders as a matter of political judgment, not rational economic theory building. The accommodationist response, although economically unsound, is valuable. It provides a critique of the traditional model and provides a base for the development of a more responsive nuclear regulatory theory. The more responsive theory is developed in chapters six and seven.

Note the central similarity among these dualities —each pair is theoretically mutually exclusive. Conceptually, these dualities do not permit a simple balancing of interests. In practice, the dualities are blended, and the blending produces unsatisfactory accommodations. Indeed, the indeterminacy and lack of theory that result from the blending and accommodations evince a change in regulatory paradigms.[5] It is because of, not in spite of, the blending that the need for a new theory is recognized, and it is from the accommodations that the new theory and new model will emerge.

Note also that the dualities describing nuclear power regulation can be dichotomized into traditional and postindustrial models. The traditional model is one in which a centralized nuclear policy is actively promoted by a joint industry-government venture. The policy is accepted by the consuming public on the belief that larger-scale, high-technology electricity will be made available safely and cheaply. The traditional policy is so uniformly adopted that a nuclear market is created by the state to foster the policy.

The traditional concept of electricity regulation posits that increasingly large, central power station-generated electricity is the preeminent form for delivery. Nuclear power plants too comfortably fit into this scheme of things. Regulators encouraged and regulations provided an incentive for the construction of nuclear power plants. This observation was made in the last chapter. The underlying theme of the traditional model is capital investment for capital expansion. The model was based on the idea that energy production was linked to economic growth, and large central power stations were seen as the perfect counterpart to large industrial growth and development.

The postindustrial model shares none of these characteristics. A pronuclear policy is not universally accepted. Instead, nuclear policy is factious, even between industry and government. The public is suspicious of the safety of plants and knows that nuclear power is not cheap. Policy making is decentralized between the states and the federal government, and among the states. Large-scale, high-technology power is no longer the only source of electricity. Finally, there is no faith in a state-dominated nuclear market. Rather, the private market dominates the question of whether a plant will get built, and the private market will be used to set prices.

Although the energy-GNP link has been questioned,[6] we need not dispense with the assumption that increased energy production means an increase in economic growth. Instead, we can accept this assumption and ask an entirely different question: If large utilities were part of the country's growth and development, does that relationship hold in a postindustrial world? The relationship between energy production and economic productivity is a question of scale.[7] Assuming that it is desirable to produce and deliver energy according to the scale on which it will be used, then we must ask whether the legal rules that regulate the production, delivery, and pricing of energy observe the principles of scale just mentioned. Therefore, as a matter of sound nuclear policy, a choice must be made between a traditional/capital expansion model or a postindustrial competitive model.

These two models are mutually exclusive, and the country is moving from the traditional to the postindustrial model. The country is now between models, and here the signs of transition will be explored.

The central, emblematic concern of this chapter is the conflict between shareholders and ratepayers when a plant is canceled. That is an ever-present distributional conflict of a zero-sum nature and, therefore, has economic and political consequences.[8] Further, the resolution of this conflict influences and is influenced by how we think about future nuclear policy making. The rudiments of this conflict can be seen through a fictitious dialogue between a utility and consumers.

> RATEPAYER: "You canceled the plant and want me to pay for your failure. I refuse. I shouldn't pay for what I don't get."
>
> UTILITY: "Let's take a more systemic look at the problem.
>
> As a utility I have a legally imposed service obligation, and our commitment to build the nuclear plant was correct and prudent when made and was made in fulfillment of that service obligation. Consider the investment as one not entirely made in a specific project, but rather as one made in satisfaction of our service obligation."
>
> RATEPAYER: "I'd accept your systemic, service obligation argument, but for one thing, your argument gives me little or no protection against overinvestment. There's no check against your continuing to dump money into construction projects if there's an effective guarantee that the costs, even for failed projects, will be passed through to me."
>
> UTILITY: "No, you're wrong. We don't 'overinvest' or 'underinvest.' We can only make prudent investments, and when we decided to build, the investment was prudent. Therefore, we should be compensated. The project might be our mistake, but it's not our failure. If anything, the cancellation is caused by regulatory failure, not managerial failure."

This short dialogue is a reconstruction of a more complex theoretical debate. The debate is about fault, blame, and recoupment. Ratepayers blame utilities, who blame regulators. Regulators fault utilities, who fault the market, and shareholders seek recoupment from ratepayers. The debate also contains mutually exclusive views of regulation. The dichotomies posed are between managerial or regulatory failure, between project-specific or systemic analysis of the cancellation problem, and about the choice between two regulatory standards known as the "used or useful" and the "prudent management" tests. These are the standards applied by regulators in determining how much and against whom the costs of canceled plants must be assessed.[9] These tests are notably indeterminate, and they are not applied with uniformity. Sometimes one test is used to preclude passing costs to ratepayers; sometimes the other test is used to justify sparing shareholders; sometimes both are used; other times hybrids develop.[10]

The fact that regulators have conflicting tests and perspectives to choose from and use indicates that the regulatory system has no adequate standard with which to resolve the conflict between ratepayers and shareholders. Instead, the choice of test or perspective acts as a justification for the result the decision maker wishes to reach.[11]

Instead of ironing out a workable theory, decision-making bodies such as PUCs, courts, and legislatures have fashioned a rough justice by accommodating conflicting claims. This accommodation is bottomed on an appeal to equity, because the existing regulations have no sound economic model providing an efficient solution. The economically sound policy is one that tracks the real market of the late 1970s and 1980s rather than the artificially created market of the 1950s through the mid-1970s. An efficient solution would price electricity based on a more competitive rather than a monopolistic market, thus assuring the proper production and allocation of resources.

Although the accommodationist response is unsound economically, it is politically understandable and defensible. The accommodationist response is politically understandable because this period is one of transition. It is defensible as long as it leads to more responsive and more responsible deicison making. That will occur as nuclear regulation changes with market changes, and as regulation moves from the traditional to the postindustrial model. If nuclear power weathers the transitional storm to emerge again, it will enter a world in which the nuclear option is chosen by a market in which financial risks and liabilities are allocated efficiently, and by a politically accountable regulatory regime in which safety and technological risks and liabilities are allocated equitably.

STATE REGULATION

In *Pacific Gas & Electric*,[12] the radiological/nonradiological distinction was used to apportion nuclear policy-making power between state and federal governments. Aside from this device, states have always had the power to set electricity rates through their PUCs. The exercise of that power is established by various statutes and regulations that require rates to be "just and reasonable" and by state and federal constitutional prohibitions that prevent the taking of private property for public use without just compensation.[13] This legal scheme has the intent and effect of balancing the interests of ratepayers and shareholders.[14] Nevertheless, the balancing of interests should be done according to some sound theory. That is not now the case, as uniformity among states is lacking, although a general approach can be noted.

While PUCs have initial responsibility for ratemaking, they do not have exclusive authority. State legislatures and courts supplement, check, and balance PUC decision making. Not surprisingly, during a transformative stage of policy development with the complexities and uncertainties of nuclear power, conflict among decision makers is inevitable.[15]

State regulators are faced with deciding who pays for the mistaken investment in an unnecessary nuclear plant. The question "Who pays?" has three subparts. First, how should the costs of a canceled plant be treated? Second, how should investment in excess plants be carried for ratemaking purposes? And, finally, while a plant is being constructed, who should pay the carrying charges? Put more simply, the three most pressing and controversial issues before state regulatory bodies are plant cancellations, excess capacity, and construction work in progress (CWIP).[16] Ultimately there are only two or three groups of likely cost bearers for these expenses: ratepayers, shareholders, or taxpayers. The primary allocation is being made between ratepayers and shareholders, while taxpayers are tapped to the extent that a utility takes a tax write-off for losses on nuclear investments.

The resolution of these issues challenges the traditional model of nuclear regulation. Further, the development of a resolution requires a deep and serious rethinking of the nature and purposes of public utility regulation. Society is now experiencing and will continue to experience this rethinking, while legislatures, courts, and PUCs decide where and how to spread costs and distribute cash invested in nuclear power. These state decision makers are deciding if and when customers should pay for electricity to be generated from a plant being built. Although each state's public utility laws and regulations have their own peculiarities, nuclear cost-allocation decisions can be grouped into three fairly broad and distinct classifications.[17]

1. States Protecting Consumers

The few states that disallow cancellation costs in a utility's rate base premise the disallowance on a simple quid pro quo: You pay for what you get. The corollary is that ratepayers need not pay for canceled plants that produce no electricity. On the surface, this posture appears to be solely based on consumer protection. Protectionism is one important, but not unproblematic, aspect of cost allocation. Another aspect is economic efficiency. If utilities are compensated for useless plants, then consumers become guarantors of utilities' wasteful investment, and utilities are shielded from normal market risks.[18]

The Ohio Supreme Court was the first state high court to refuse to allow the recovery of abandoned nuclear plant costs from ratepayers by not allowing a nuclear generating unit to be included in the rate base until used and useful. In *Office of Consumers' Counsel* v. *Public Utilities Commission of Ohio*,[19] the Ohio court refused the request of the Cleveland Electric Illuminating Company to amortize the cancellation costs of four nuclear power plants and, in so doing, reversed the Public Utilities Commission of Ohio (PUCO). The court based its decision on a state statute allowing a utility to recover in its rates "the cost to the utility of rendering the public utility service for the test period."[20] The PUCO interpreted this language so that "an expenditure by a utility can be considered a cost of rendering the public utility service if it fails in fact to achieve its intended purpose . . . [if] the expense was reasonably calculated to provide [future] utility service at a reasonable cost."[21] The commission found that the decision to build and the subsequent decision to cancel the plants were prudently made, so the costs should be amortized. However, the Ohio Supreme Court interpreted the statute to preclude the extraordinary costs of a terminated nuclear power plant from being paid by consumers. Instead, the statute was held to account for the normal, recurring expenses of the utility. Had the projects been completed, their costs would have been allowed in the rate base, because investors are entitled to receive a fair and reasonable return on used and useful property. Since the plants were never used and useful, the costs could not be imposed on ratepayers.

Although the Ohio decision was based on statutory language, the court also took issue with the PUCO's policy considerations. The commission argued that disallowing abandoned plant costs would tempt utilities to complete imprudent projects, so they would ultimately earn a return on their expenditures. To an extent, that is true. The rate formula rewards completed capital projects. However, the court noted that since the PUCO had the power to investigate the management of the utility, it also had the duty to disallow rate base treatment and any claimed operating expenses associated with imprudent investments such as imprudent completion of an unnecessary project.

The disagreement between the PUCO and the state supreme court is a replay of the fictitious dialogue and demonstrates the competing policies behind the allocation of cancellation costs. The PUCO chose to evaluate cancellation costs as a question of prudent or imprudent management. The court, however, applied the "used or useful" test. The choice of test was crucial. If the "used and useful" test is adopted, then generally, shareholders will absorb cancellation costs, because there is no plant on line. If the prudent-management test alone is chosen, then generally ratepayers will pay.[22] Still, utilities are not without avenues of relief. They can convert from nuclear to coal as long as the decision to switch is prudently made. Or, the utility can request a higher rate of return in order to attract new investment.

In a later case, the Ohio Supreme Court rejected the notion that disallowing recovery of cancellation costs was an unconstitutional taking of private property. Relying on a United States Supreme Court opinion, the court noted that in computing a rate base "postponement of . . . profit until the state of imminent or present use is not an act of confiscation, but a legitimate exercise of legislative judgement."[23] The Ohio court decided that investor and consumer interests had been properly balanced, because "investors are assured a fair and reasonable return on property that it determined to be used and useful, . . . plus the return of costs incurred in rendering the public service, . . . while consumers may not be charged 'for utility investments and expenditures that are neither included in the rate base nor properly categorized as costs.'"[24] The Ohio court did not accept the argument that disallowance would seriously disadvantage Ohio in capital markets, stating that that was a concern for the legislature. However, the Ohio Supreme Court later found that the PUCO had the discretion to take the disallowance into account in determining a fair and reasonable rate of return: "The question whether a decision of this court may have so increased the perceived risk to investors as to require a higher rate of return on common equity is one the commission may consider as a factor in its decision."[25]

Wyoming also denied recovery from ratepayers for the costs of the canceled Pebble Springs and WPPSS-5 plants because they were not on line.[26] By statute, Wyoming requires the utility commission to consider "the property and business of any public utility, used and useful for the convenience of the public."[27] The commission held that the risk of loss to public utility investors is not lessened by the fact that a public utility incurs a service obligation. The Wyoming Supreme Court affirmed this decision because the commission never approved the construction projects, and thus the consumers never had a chance to consider and accept the risks involved.[28] The utility argued that the costs should still be recouped as prudent, reasonable, and proper expenses of operation. After noting that since the projects were never completed, because they did not meet the "used and useful" requirement, the court rejected the operating-expense argument, because cancellation costs simply did not fit the traditional definition of operating expenses. The court pointed out that the utility would never have claimed these costs as operating expenses had the projects been completed. Instead, the utility would have included them in the rate base. The Wyoming Supreme Court imposed the losses on stockholders because "the decision to embark upon the project was made by their representatives who had an opportunity to calculate the risk and be governed by such calculation." If the consumers bore the costs, then there would be no risk at all for the stockholders, which would encourage the utility "to venture into activities having a very small chance of economic success with the knowledge of no loss to it should the activity fail and of great gain should the small chance of success occur."[29]

Oregon does not allow recovery of abandoned plant costs because of the enactment of "Ballot Measure 9," which provides:

> No public utility, shall directly or indirectly, by any device, charge, demand, collect or receive from any customer rates which are derived from a rate base which includes within it any construction, building, installation or real or personal property not presently used for providing utility service to the customer.[30]

After passage of this act, two utilities requested rate relief to compensate for cancellation losses attributed to Skagit/Hanford and WPPSS-5. According to the Oregon commission, Ballot Measure 9 clearly proscribed any recovery of abandoned-plant costs from ratepayers. However, the Oregon commission refused recovery of only those costs incurred after the passage of this statute, because at the time "utility management was fully aware that its investment was a risk and the previous risk-sharing relationship between investors had been altered."[31]

The Montana Public Service Commission (MPSC) disallowed recovery for the abandoned Pebble Springs and WPPSS-5 nuclear projects based on both the state statute and policy considerations.[32] The state statute provided that "the commission may, in its discretion, investigate and ascertain the value of the property of every public utility actually used and useful for the convenience of the public."[33] The MPSC said that it could perceive no legislative purpose in including the phrase "actually used and useful" other than to limit a utility's recovery of investments in property to those investments in property that actually provide service.

The MPSC rejected the utility's attempt to limit the statute's application to rate base determinations. The utility sought to recover its total investment by being allowed to amortize costs over five years, plus recover a carrying charge equal to the utility's rate of return. According to the MPSC, "The only difference to the company between one scenario where the plants are completed and are used to provide electric service and given rate base treatment and a second scenario where the plants are abandoned and are given the treatment proposed by the company, is that the company would receive its total investment faster in the second instance. If anything, the company is better off if the projects fail and have to be terminated than it would be if they are successfully completed and provide service."[34] Consequently, the MPSC refused to allow the utility to circumvent the statute by indirectly receiving the same or better benefits as rate base treatment. The MPSC rejected the argument that total disallowance of cost recovery would constitute a taking of private property in violation of both the United States and the Montana constitutions. The MPSC stated that neither a "taking" nor a "use" of private property had occurred, since the utility retained full control and use of the projects unrestrained by either the commission or the ratepayers.

The MPSC supported its decision with its own policy considerations. First, the commission disregarded the argument that a regulated utility is not on the same footing as a nonregulated business because a utility's obligation to serve the public restricts its activities. Instead, the MPSC stressed that investors in regulated utilities experience advantages because their investment is less risky than most investments in nonregulated businesses. Second, allowing recovery for all losses improperly expands the monopoly advantage to the point where investments would be risk-free. Furthermore, by adopting the "used and useful" standard, the legislature put the utility investor on notice of the possibility that there may be no return on investment if the project does not become operational and beneficial. Finally, the MPSC found that even in the absence of the "used and useful" language, it "would nonetheless not allow recovery from the ratepayers . . . after balancing the interests of the utility and the ratepayers." The MPSC felt that it would be "patently unfair" to force ratepayers to pay for canceled projects "which they had no part in conceiving, which failed through no fault of theirs and which will never benefit them."[35] Since investors, through the management they select, indirectly conceive and pursue projects, the MPSC decided that it was fair for them to bear the cost of failure.

Likewise, the Arizona Corporation Commission refused recovery for cancellation costs of Palo Verde Units IV and V. Among the reasons cited by the Arizona commission were inadequate justification by the utility for the expenses, the unusual, and nonrecurring, nature of the expenses, and the fact that "the planning of construction of new units is a management function under control of the stockholders . . . [who] should therefore bear any cost related to such cancellations."[36] However, the Arizona commission was in part displeased, because the decision to cancel "was triggered by regulatory or other conditions in the state of California," and "prudent utility management should have negotiated appropriate contractual provisions to guard against and/or recover for such a contingency."[37]

New Hampshire has joined the states that disallow cancellation costs in the rate base. The New Hampshire legislature adopted an "anti-CWIP" statute, which prevents any construction work in progress from being included in the rate base until the plant is completed.[38] The New Hampshire Supreme Court had interpreted the statute to preclude utilities from including in the rate base those cancellation costs incurred after the passage of the statute, even though the strict language of the statute focused on CWIP.[39] Curiously, the Missouri Supreme Court interpreted a statute modeled after the New Hampshire anti-CWIP legislation to differentiate between CWIP and cancellation costs and to allow the inclusion of cancellation costs in the rate base.[40]

Indiana does not permit a pass-through of abandonment costs to ratepayers even in the face of a statute that arguably permits the PUC to make accommodations in setting rates. The Indiana Public Service Commission (PSC) allowed a partial recovery of the canceled Bailly N-1 nuclear plant through a fifteen-year amortization period. The commission favored an equitable sharing of $190.7 million between the ratepayers and common stockholders because the PSC gave its blessing to the project, and, as a government agency, it must act in the public interest. Furthermore, the PSC allowed the amortization because of the utility's service obligation. The utility characterized the cancellation costs as an "extraordinary cost of service loss incurred in an effort to provide energy for its system." The commission was reversed on appeal when the intermediate appellate court and the state supreme court rejected the systemic argument and applied the "used and useful" statutory standard for establishing reasonable rates, thus disallowing the costs. The court applied the "used and useful" test in denying recovery as the way to balance investor and consumer responsibilities:

> In the competitive market, investors contribute capital which is employed to produce a product. Consumers purchase the product, and the purchase

> price includes a reimbursement for the capital contribution of the investor plus a profit to compensate the investors for the risk they assumed. In the market under consideration here, the utility is granted a monopoly. Utilities are regulated in order to protect the consumers from the abuses of monopoly, i.e., artificially high prices. The statutes which govern the regulation of utilities and which grant the PSCI its authority and power provide a surrogate for competition. See *Public Service Commission v. Indiana Bell Telephone Company*, (1955), 235 Ind. 1, 130 N.E.2d 467. I.C. § 8-1-2-1 and IC § 8-1-2-4 insure that the responsibilities of utility investors and consumers are commensurate with the responsibilities of investors and consumers in a competitive market.[41]

Temporary cancellations can be treated similarly. After TMI, Unit No. 2, the damaged generator, was removed from the utilities' rate base. At the time of the accident, Unit No. 1 was down for refueling, and it remained down until 1985. The Pennsylvania PUC reduced the utilities' rates as a result of the downtime, because Unit No. 1 was not used and useful. The utilities filed complaints that the reduced rates were unjust and unreasonable, and they sought increases. The PUCs granted rate increases that allowed only the inclusion of start-up costs for Unit No. 1 but not for the downtime, and the utilities appealed. The Pennsylvania Supreme Court rejected the utilities' argument that rates had to be set at a level that preserved their financial integrity:

> If the impact of diminished financial integrity were shifted from utility companies to the consumers, as would be the case if the utilities were regarded as having a constitutionally guaranteed right to rates which would preserve their financial integrity, elevating their rates above those levels that would otherwise be regarded as providing a "just and reasonable" return on assets utilized in the public service, the result would effectively circumvent the longstanding principle, discussed supra, that property included in a rate base must be "used and useful" in the public service. Specifically, rate bases which had been determined through the exclusion of non-useful property would be rendered superfluous because the resulting rates would have to be adjusted upwards, whenever the financial integrity of the enterprise was in peril, and the effect upon rates would be the same as if portions of the non-useful property, such as the Three Mile Island nuclear reactors, had not been excluded from the rate base in the first instance.[42]

Also, in July 1976, Pacific Gas and Electric shut down its Humboldt nuclear plant for refueling and seismic modifications. At the time of its 1979 rate hearing, the plant was still not on line, and PG&E could give no convincing estimate when the plant would return. The California PUC, therefore, took the plant out of the rate base while it was shut down, because the plant was not used or useful.[43]

This group of cases present interesting trade-offs. More significantly, these opinions cannot be taken at face value. Both Ohio and Pennsylvania, as examples, will not allow ratepayers to pick up cancellation costs. However, both states have also engaged in negotiated settlements in which some costs were apportioned to ratepayers. Ratepayers agree to pick up some costs and avoid later rate shock or higher interest charges, and utilities reduce risk and acquire some financial stability.

Those courts and commissions that strictly adhere to the "used and useful" standard can preclude a utility from giving rate base treatment to even prudently incurred cancellation costs. The states that adopt a consumer protection stance do so for two important reasons. First, by statute only used or useful plant costs can be passed on. Second, these states explicitly address the fundamental issue of risk involved with cost allocation. Costs should be allocated to risk bearers in proportion to their risk assumptions. Ratepayers do not embark on new construction projects, utilities do. Utilities reap rewards if the plants are successful; accordingly, they should suffer the losses of failure. This risk allocation theory requires imposing risks on the parties responsible for making the choices. This theory, however, does not convince all regulatory authorities. It is not inevitable that PUCs require on-line status for plants before the utilities recover some costs. Instead, PUCs can and have been more generous to utilities.

2. States Protecting Investors

To a large extent, New York is on the opposite end of the spectrum from Ohio, Montana, and like states. The New York Public Service Commission (PSC) has "a practice of allowing full recovery of all sunk costs, including carrying charges, irrespective of the relative benefits that may have flowed from the abandoned or uncompleted project." Plant construction is not considered a "joint venture" between ratepayers and utilities, because "the statutory obligation imposed on utilities to serve their customers undoubtedly has a significant impact on management decisions to undertake the construction of large generating plants."[44] The PSC, thus, rejected the idea that the costs of the abandoned Sterling nuclear project should be allocated between ratepayers and shareholders according to the benefits each group would have received had the project been completed.[45] The PSC agreed with the utilities that ratepayers would have been the primary beneficiaries of a completed Sterling plant, and any benefits to shareholders would be short-lived because of regulation.

Once the commission decided that ratepayers should properly bear the costs of cancellation, it then turned to the question of how to structure recovery. There are two basic choices and several variations. Costs can be recouped with rate base treatment, which thus gives shareholders a return on failed investment. Or costs can be recovered as expenses and amortized over a period of years. The variations come in the length of the amortization period and in the treatment of the unamortized balance. The PSC balanced the impact on consumers associated with alternative amortization proposals with the effect on the financial integrity of the utilities. Recognizing that cash flow improves with a shorter amortization period, the PSC granted amortization periods of three to ten years to various utilities, and it also granted a carrying charge on the unamortized property losses based on the utilities' costs of capital.

The benefit that shareholders may enjoy from the PSC's Sterling decision are not equally enjoyed by Shoreham's Long Island Lighting Company (LILCO) shareholders. Even though the PSC can pass through costs for plants that are not on line, the PSC cannot pass through costs imprudently incurred. The PSC has ruled that $1.395 billion of Shoreham cost overruns must be recovered from LILCO shareholders, not from ratepayers, because of imprudent management.[46]

The Connecticut Department of Public Utility Control (CDPUC), like the New York commission, has placed the burden of the loss on consumers. The CDPUC allowed United Illuminating Company to recover $14.7 million of costs for Pilgrim Unit 2 by amortization over a two-year period but allowed a ten-year amortization period for a $200 million investment in another plant. The department rejected arguments that costs should not be recovered because the unit had never been "used and useful," that the company's financial integrity was not endangered, and that the utility was bound by earlier representations that it would bear the loss if the unit was not completed. Instead, it reasoned, United Illuminating acted reasonably and with the department's approval. Therefore, if the utility was satisfying the obligations of law, then it should not be penalized. In a later case, the CDPUC treated cancellation costs differently, and a utility was allowed full recovery of expenses except for carrying charges on the unamortized portion.[47]

Courts and commissions walk a tightrope between ratepayers and shareholders. Even when a decision maker acts decisively on one issue, such as allowing recovery of canceled plant costs, the decision maker can penalize a utility for poor management. There is a great deal of slippage in the economic rationales given in both sets of cases. The consumer protectionist states argue that no plant means no charge, because consumers bear no risk for failure. Investor protectionists argue that references to competitive risks are inapposite, because, unlike the free competitive market, utilities must provide service by law, and they cannot invest in other products or stop production. The slippage between the two arguments is the fallacious belief that the choice between protecting shareholders or ratepayers must be based on the traditional model of regulation. The traditional model requires rates to be derived from rate base, and risk is accordingly allocated by regulators. Another choice exists between the traditional model, which has generated a series of uncomfortable accommodations, and the postindustrial model, which ignores a rate base method and sets prices according to a more competitive environment. Thus, risk is allocated more by the market than by regulators. In the next group of cases, decision makers more explicitly and directly accommodate competing interests.

3. States Balancing Interests

The majority of states divide the burden of cancellation costs between investors and ratepayers. The policy position behind the allocations is easy to state, but the precise theory on which the allocations rest is not so easy to identify. Consumers want reasonable rates and reliable service; utilities want financial health; and investors want a return on investment.

Neither of the two previous subsections was based on an irrefutable argument. Consumer protection may harm utilities, and shareholder protection requires subsidization by ratepayers. The natural tendency is to develop an accommodationist position that attempts to protect the various interests. The accommodationist attitude is a natural reaction to a transition. Accommodation attempts (and succeeds at) making decisions without a solid theory, because decisions must be made in the short term. In a pluralistic democracy and in an activist state, the accommodationist response is understandable and legitimate, again in the short term, precisely because decisions are made. The accommodationist response is inadequate in the long term because it is a response (action) without a theory. As a matter of clarification, the short term should last no longer than the briefer of two periods: first, when all current nuclear plant cancellation decisions have worked their way through the regulatory state; or second, when a sound theory is developed. Finally, that theory must be based on the lessons of the transition; otherwise the accommodations go for naught.

The Massachusetts Department of Public Utilities (MDPU) directly confronted the nature of accommodation:

> The primary mandate that rates balance the interests of consumers and investors must be applied here based upon our judgement of the appropriate factors that affect such a balancing. It seems indisputable to us that no mathematical formula, including one that evenly divides dollar losses, can properly and logically effect a meaningful balance of interests. After considerable review, we have concluded that the factors which properly bear on the allocation of the loss at issue are the following:
>
> a. The prudence of the company's actions throughout the history of the project;
> b. The equity and fairness of any proposed allocation; and
> c. The necessity of adjusting the financial impacts of any allocation to ensure the adequacy of future service.[48]

The Pilgrim II nuclear power plant was canceled by the board of directors of Boston Edison on September 1, 1981. Boston Edison, the lead participant and 59 percent owner of the project, invested over $278 million, an amount equal to about two-thirds of the utility's permanently invested capital. Because of the magnitude of the loss, the MDPU took a de novo approach in allocating the loss. Although the "used and useful" standard precluded rate base treatment for the plant, this standard, according to the MDPU, was never meant to preclude recovery entirely. Instead, the MDPU adopted the prudency standard [49] and subsequently disallowed recovery for the period after July 1, 1980, the point at which project uncertainty had become unacceptably high. The MDPU decided to amortize over a thirteen-year period the direct costs and the debt component of AFUDC. The MDPU also allowed a carrying charge on the unamortized portion of the prudent expenditures but did not, however, allow the unamortized portion of the plant into the rate base on the premise that the utility would regain expenses, but at a lesser rate than its allowed rate of return.

The MDPU reasoned that if the utility was unregulated, the possibility of an extraordinary loss would place the company at an extraordinary risk. Investors in such "speculative" ventures receive speculative returns. Since the regulated utility would never be allowed returns commensurate with the extraordinary loss of Pilgrim II, forcing the utility to absorb the losses would inadequately compensate investors. Ultimately, ratepayers would "pay a high price in terms of both extravagant compensation for new capital and an unavoidable service deterioration reflecting the scarcity of reasonably priced capital."[50]

The Massachusetts Supreme Court upheld the MDPU's decision as within the department's discretion, noting that "investor confidence is not an insignificant element in utility regulation," and "a judgment that the adverse consequence of disallowing any recovery would be a serious threat to the company's financial integrity, and indirectly to its customers, was warranted."[51] The court also found that the "fact that Pilgrim II never produced electricity, never became used and useful, does not prohibit recognition of its prudent costs in ratemaking."[52]

There were strong dissents in the Boston Edison case at both the department and state supreme court levels. Commissioner Sprague argued that costs should be borne solely by investors, because "it is a fundamental business principle that the owners of a corporation benefit when a corporate strategy creates profits, but bear the burden when funds are expended on ventures which fail."[53] Because of the long-established rule in Massachusetts that property must be "used and useful" to be included in the rate base, combined with the Massachusetts policy of not charging ratepayers for CWIP, the commissioner agreed with the attorney general's argument that "it would be illogical to permit cost-of-service treatment for construction work *no longer* in progress and completely abandoned."[54] Judge Liacos of the Massachusetts Supreme Court stated that the purpose of regulating utilities is "to ensure to the public the availability of service at reasonable cost from private industry protected from the risk of competition." Instead of protecting the public, the majority's decision required the consuming public "to rescue the utility from its mistakes, even though the consumers bear no responsibility for the company's mistakes and derive no benefit from its acts."[55] Dissenting opinions from legal rulings are not novel. They do betray the existence of doctrinal conflict during the developmental period of a legal rule.

The Maine Public Utilities Commission has adhered to a policy of allocating the costs of canceled plants between shareholders and ratepayers by allowing amortization without rate base treatment of only that part of the investment that does not constitute capitalized AFUDC. According to the commission, "when a plant becomes used and useful, the shareholder is ordinarily rewarded for his risk by being allowed both a return of AFUDC, through depreciation, and on AFUDC, when it is capitalized and included in rate base."[56] In the case of a canceled plant, however, AFUDC represents the carrying costs of a plant that will never provide service. The PUC then decided that "a reasonable balancing" of the burden of the canceled plant required shareholders to shoulder all carrying costs. The PUC refused to disallow only the equity portion of AFUDC, because the "equitable share of risk that should be borne by investors is the risk of loss of the expected return on the investment," which includes the return on debt as well as on equity.[57]

The Maine PUC policy is that principal amounts originally invested in the project should be returned to investors by ratepayers, so that risks involved in investing in construction programs would be reduced. On the other hand, reducing the investment risk "does not go so far as to require a return of the carrying costs that represent, in part, an allowance for investor risk itself."[58] Since AFUDC represented these carrying costs, the PUC denied its recovery.

The Illinois Commerce Commission allowed a sharing of the costs of the abandoned Calloway II plant to accommodate the interests of both shareholders and consumers. The commission allowed amortization of determinate costs, including accumulated AFUDC over a five-year period, but did not allow rate base treatment for the unamortized balance.[59] The commission rejected an argument by the attorney general to disallow recovery of accumulated AFUDC, which represented a reward to investors for bearing their portion of project risk during construction.

In 1982, the North Carolina Utilities Commission allowed the amortization for Carolina Power and Light (CP&L) of the abandoned Harris Units 3 and 4 over a ten-year period, and allowed recovery for the carrying charges associated with long-term debt.[60] The commission allowed the unamortized balance supported by long-term debt to be given rate base treatment because it reasoned that CP&L had a legal obligation to pay carrying charges, and also because long-term debt holders have little direct impact on management. The commission asserted that "CP&L's preferred and common shareholders who control management through its elected board of directors should not be permitted to receive a return on moneys invested by management in a plant that was subsequently canceled, even though the initial decision to build said plant and the ultimate decision to cancel same were clearly reasonable and prudent. . . ."[61] According to the commission, this decision imposed 73 percent of the after-tax capital costs on the shareholders and 27 percent on the ratepayers.

Distinguishing between debt and equity has superficial appeal. Bondholders should be protected more than shareholders, because shareholders are the owners of the utility, and bondholders are only lenders. The argument makes less sense when the realization is made that both bondholders and shareholders assess risk based on the same information about capital structure, financial risk, and liability. What the post hoc distinction between debt and equity does is to realign risk and liability after investments have been made.

The North Carolina Utilities Commission departed from its former position of granting only partial recovery by allowing full recovery of the costs of North Anna 4. The commission stated that it wanted to "avoid penalizing stockholders as a result of prudent management decisions." In a subsequent decision, the commission eliminated the inclusion of the long-term debt in the rate base, because relevant statutes allow a utility to earn a return on property only if it is "used and useful" or if it meets the CWIP statutory standards.[62] According to the commission, this result comported with its "ultimate responsibility . . . to fix rates which are fair and reasonable both to the utility and to the consumer."[63] The commission allowed a ten-year amortization, as suggested in a Department of Energy study,[64] but disallowed rate base treatment for the unamortized balance. North Carolina's rules also give favorable treatment to CWIP, as long as construction is in progress and cancellation is doubtful.[65]

The Washington Utilities and Transportation Commission also endorses a sharing of costs between investors and ratepayers. However, it refused to allow one utility to amortize losses for the canceled Pebble Springs project and instead raised the utility's rate of return.[66] The commission reasoned that since Oregon and Wyoming had not allowed recovery from ratepayers for the terminated project, the investor-perceived risk would be greater regardless of any actions by the Washington commission, so the rate of return would have to rise. Thus, to allow amortization of the costs of the terminated project would "constitute a double recovery of the Washington ratepayers' portion of investment."[67] However, in another case the commission allowed a different utility to amortize the Pebble Springs losses over a ten-year period and denied rate base treatment for the unamortized portion.[68] This sharing of fiscal responsibility, according to the commission, preserved the utility's ability to render public service at reasonable rates. The commission also refused a proposed debt-equity exchange to recover the costs from WPPSS-5, because it would be improper to set off gains with losses when full recovery was denied. By statute, Washington also precludes the inclusion of CWIP in the rate base.[69]

The West Virginia Public Service Commission allowed the Virginia Electric and Power Company to amortize losses of the canceled North Anna Unit

4 and Surry Units 3 and 4 for ten years. Rate base treatment of the unamortized portion, causing "a sharing of the risk of the project between customer and company," was "a reasonable approach since neither group was responsible for the fate of the project."[70]

The Virginia State Corporation Commission also allowed a ten-year amortization of cancellation costs incurred by the Virginia Electric and Power Company for the North Anna Unit 4. The Virginia commission, like that of West Virginia, also denied rate base treatment of the costs. The commission reached this decision because the property was never "used and useful" and because rate base treatment would severly penalize ratepayers, who would never benefit from the plant.[71]

The New Jersey Board of Public Utilities endorsed a sharing of abandoned-plant losses by investors and ratepayers by amortizing Hope Creek II losses over a fifteen-year period instead of the five-year accelerated period suggested by the utility. The board did not give the unamortized portion rate base treatment, since the property was not "used and useful," thus denying the company's request for carrying charges. In contrast to the Massachusetts Department of Public Utilities, the New Jersey board found that granting carrying charges would give the utility the equivalent of rate base treatment and would improperly allow investors to earn a return on property that was never used or useful.[72]

Idaho favored ratepayers by granting a fifteen-year amortization period to the Washington Water Power Company regarding the cancellation of the Skagit/Hanford plant.[73] The PUC reasoned that a fifty/fifty sharing of costs could be accomplished through the lengthy amortization of 48 percent of the utility's requested costs.

The Vermont Public Service Board allocated the losses from the Pilgrim Unit 2 abandonment by the Central Vermont Public Service Corporation by amortizing these losses over a ten-year period with no rate base treatment for the unamortized balance. The board engaged in a "risk return relationship" analysis and stated:

> [I]f all costs of every failed investment are to be recovered in rates, then there is no utility risk, and the required return drops dramatically to a very low risk-free rate such as a passbook savings rate. On the other hand, if shareholders perceive when they purchase stock that there is some risk to their investment because all of the costs associated with an abandoned plant will not be collected, then they demand a higher rate of return (which they collect before the plant is abandoned) and thus will have been compensated for any "loss" occasioned by regulatory treatment.[74]

Since the majority of states and the Federal Energy Regulatory Commission do not allow rate base treatment for plants not used or useful, the board reasoned that investors should have discounted this regulatory treatment already in the rates of return demanded by them.

Although states articulate and apply different rules and tests to allocate cancellation costs, there is a common theme running through the decisions: The nature of public utilities as regulated monopolies is being reevaluated. As the Vermont Public Service Board put it:

> [How] to treat the loss of an abandoned plant presents one of the fundamental regulatory paradoxes. Regulations' role is to supply the discipline of the marketplace to a monopoly that provides an essential service. If an unregulated company subject to competition makes a bad investment, the company is unable to pass that along to its customers and its shareholders suffer from decreased earnings or from the ultimate economic fate of bankruptcy. On the other hand, an unregulated company can also earn "surpluses" above the average return in the industry which will cushion it in hard times. However, if a regulated company providing an essential service makes an investment that is later abandoned, to disallow it completely and reduce earnings only raises the cost of debt and equity to the company, and this cost is passed along to customers in the long run.[75]

As the United States Supreme Court has stated, "The nature of government regulation is such that a utility may frequently be required by the state regulatory scheme to obtain approval for practices a business regulated in less detail would be free to institute without any approval from a regulatory body."[76] In allocating losses between ratepayers and investors, PUCs try to balance the utilities' burdens as a regulated industry against their privileges as monopolies.

Utilities argue that since costs of canceled plants were incurred as part of their service obligation, the public should share in these costs. At the time most nuclear plants were begun, the federal government was vigorously promoting the development of nuclear power. In the 1970s, the energy crisis and the resulting need to become less dependent on foreign oil increased reliance on nuclear power. Some PUCs are reluctant to deny financial assistance to a utility after it finds that the decision to build the nuclear power plant was a prudent effort to fulfill the utility's service obligation to the public, particularly when nuclear power was so promoted. Other PUCs reason that cancellation costs are just a "risk of doing business." Both arguments are off point. As regulated firms, utilities are neither competitors in the market nor the sole preserves of government. They are a hybrid for which a clear economic model and standard financial analyses do not fit squarely.

Investor risk is another factor debated by PUCs. Utilities argue that the return to investors can never be as great as the return earned on more speculative ventures. Nevertheless, utilities fear that investors will stay away if consumers do not bear some of the losses, because the costs of raising money for further construction will be prohibitively high. However, most PUCs are concerned that if some costs are not borne by investors, then investors' risk would impermissibly be reduced to zero. Lastly, the threat of bankruptcy and possible interruption of utility services has persuaded some PUCs to find ways to give utilities financial relief. To prevent bankruptcy, PUCs will either allow recovery of some or all abandonment costs or grant emergency rate hikes.

The state regulation of nuclear power is confronting the future. As the once highly centralized nuclear decision making power is being diffused, state regulation gains in importance. Traditionally, state utility regulation simply passed the cost of new plants on to ratepayers. Today, however, cancellation costs threaten to impose financial hardship on the industry. Thus, regulators are faced with deciding who should absorb these costs. Should ratepayers absorb all or some of the costs of a utility's investment decision that produces no electricity in an effort to lend financial support to the utility and ultimately themselves? Through a series of accommodations and trade-offs, the near-unanimous answer is that ratepayers will pick up some charges and shareholders other charges. However, the precise method of allocation is still being worked out. In the process, regulators must now address how best to treat, for ratemaking purposes, large investments in projects with long lead times. The answers that are being developed will directly affect the financial structure and capital budgeting of utilities. Ideally, utility investments will accurately reflect market conditions, so that productive and allocative efficiencies are realized.

These state decisions pose a seemingly intractable dilemma. Should consumers pay now for the construction of large projects that 1) may not be completed and 2) may subsidize a different generation of consumers in order to avoid rate shock? Or, should consumers pay to maintain the integrity of the system as a trade-off for the utility's service obligation? Either way somebody loses, because the risks are artificially imposed by the regulatory system rather than taken from the market. The ideology contained in the dilemma is that somehow large central power stations deserve protection from market risks. Answering this dilemma depends on a frame of reference. If the problem is analyzed from a project-specific standpoint, then canceled-plant losses are suffered by utilities. If a systemic analysis is used, then ratepayers have assumed some risk, because they are the beneficiaries of the state-imposed service obligation. Both arguments are partially right and, hence, partially wrong. The arguments are right insofar as each leads to a determinable answer. The project-specific argument favors ratepayers, and the systemic argument favors shareholders in cost-allocation decision making. Both arguments are wrong because neither alternative produces completely sound results. A decision maker has the discretion to choose either analysis as a justification for the result he or she wishes to reach. There is no meta-decisional rule that requires the decision maker to adopt one argument or the other. The project-specific/systemic dichotomy is simply another version of the mutually exclusive dualities inherent in regulatory analysis during this transitional period. This decisional indeterminacy is constitutive of the breakdown of institutionalized policy during a period of dynamic change.[77] The way to break out of the dichotomy is to move toward more market-based ratemaking rather than a cost-based method.

FEDERAL REGULATION

Federal authority over cost allocation is less pervasive than the power and authority exercised by state PUCs, but it is significant. Just as the states are not uniform in their treatment, federal regulators are no longer of one mind in their attitude toward nuclear power. The nation responded to the TMI accident with appropriate caution. But old ways die hard, and the federal regulatory structure is no less immune from change than any other institution. Specifically, the caution following TMI has given way to the promotion reminiscent of earlier times.

Several federal agencies have responsibility for nuclear power regulation,[78] although the NRC, the Federal Energy Regulatory Commission (FERC), and the Department of Energy (DOE) predominate. Each of these agencies has different responsibilities. Combined, they present a picture promotive of nuclear power.

1. Department of Energy

The DOE has the least-direct impact on the daily regulation of nuclear power of the three agencies discussed. Instead, the DOE sets the tone for nuclear policy making in its role as information gatherer for other government agencies.[79]

When the Atomic Energy Commission was reorganized in 1974 and its promotional and regulatory functions were supposedly separated, the Energy Research and Development Agency (ERDA) and NRC respectively were created.[80] ERDA was the research arm for all energy areas, not just nuclear power. In 1977, with the creation of the DOE, ERDA was subsumed by the new agency and was reconstituted into the Energy Information Administration (EIA), which is responsible for carrying out a central comprehensive and unified energy data and information program.[81] The EIA collects, analyzes, and disseminates data relevant to energy resource reserves, production, demand, and technology. The mission of the EIA is to accumulate energy data to meet near- and long-term demands for the nation's economic and social needs.[82] The DOE also has an assistant secretary for nuclear energy,[83] who is charged with the department's waste management responsibilities.

Aside from information gathering, the DOE has no direct responsibility for allocating abandonment costs. However, through its publications and official pronouncements, the department has promoted the nuclear industry.[84] Given the history of the creation of the DOE, this posture is understandable. In large measure, the department was the final embodiment of an evolving response to the Arab oil embargo of 1973 and to the inflation that preceded it. At that time, the disruption of our oil supply was perceived as the most serious threat to our national security, as well as our national energy plan, such as it was. Multiple oil price hikes dislocated our economy with their inflationary impact and skewed our energy mix by making oil- and gas-based electric-generating units too expensive to run.[85] In an effort to protect the economy from severe price shocks, President Nixon, in phases three and four of a series of wage and price controls, placed a ceiling on oil prices, which was first administered by the President's Cost of Living Council.[86] The Emergency Petroleum Allocation Act[87] was the direct response to the OPEC embargo and was intended to provide for the equitable allocation of crude oil and refined products. The act was administered through the Federal Energy Administration, which was absorbed by the DOE.[88]

The DOE was born from an oil crisis. Because of the interdependent character of energy planning, how the department treats oil affects nuclear power. The perception of energy scarcity meant that the country had to reduce its dependence on all oil, foreign as well as domestic. Power plants, together with large industrial oil consumers known as major fuel-burning installations, were targeted for oil cutbacks.[89] As plants converted from oil and natural gas, they needed other fuels, and coal and nuclear were the obvious alternatives. Consequently, the DOE was predisposed to encourage the use of nuclear power as a way of relieving the pressure on oil use.

Although the DOE has not advocated a financial bailout of the troubled industry, it has been a supporter. Under Energy Secretary Hodel, the DOE was committed to ensuring that nuclear power continued to play an important role in securing America's energy future, because nuclear power was seen as "critical" for a balanced energy mix, particularly because of its potential to displace oil.[90] To help reinvigorate the industry, the DOE supported licensing legislation designed to shorten the construction lead time, which is currently twelve to fourteen years. This proposal is intended to reduce regulatory costs.[91]

The uniform DOE message is the need for growth along the hard path. DOE projections of the need for additional generating capacity are consistent with industry figures.[92] With reduced public consciousness of the "energy crisis" in the early 1970s which gave birth to the DOE, a business-as-usual approach has been adopted by the agency. The large, capital-intensive central power station is the energy path of choice for the nation's central energy-planning agency. This policy option, coupled with a historical distrust of oil reliance, means that nuclear power is touted.[93]

2. Federal Energy Regulatory Commission

The FERC has direct authority over allocating cancellation costs. However, it has only a regional impact. The FERC is something of a bureaucratic oddity.

When the DOE was established, all of the powers of the Federal Power Commission were transferred to it, and the FPC was renamed the FERC. The FPC was an independent regulatory body, and the DOE is an executive, cabinet-level agency. In order to consolidate energy planning, the FPC was nominally placed under DOE jurisdiction; however, it retains its independent authority insofar as the FERC has final say over the matters before it. In some instances, the FERC may veto or supersede actions of the secretary.[94]

The FERC, through Part II of the Federal Power Act,[95] exercises regulatory, including ratemaking, jurisdiction over interstate sales of electricity for resale. The commission has other responsibilities, such as the power to divide the country into regional districts for voluntary interconnections and to order interconnections.[96] It has not used this authority to create a national grid, which, some argue,[97] may become a future necessity to avoid reliability problems. The FERC's jurisdiction over electricity was extended in 1978 as part of President Carter's National Energy Plan. One piece of legislation that emerged from the plan was the Public Utility Regulatory Policies Act (PURPA),[98] which was the first federal intrusion into retail ratemaking. Under PURPA, the DOE was allowed to intervene in state ratemaking proceedings, and the FERC was authorized to order interconnections, wheeling, and pooling and to direct interconnection between small power and cogeneration facilities.[99]

The FERC's expansion of power by PURPA, held constitutional by the Supreme Court in *FERC* v. *Mississippi*,[100] is significant. Electricity regulation, particularly ratemaking, is traditionally a local matter. Recall that *Pacific Gas & Electric* explicitly recognized that setting rates for nuclear-generated electricity was a local function. However, with PURPA there exists the potential for the recentralization of authority over nuclear power by moving toward uniform national ratemaking standards, and toward a national grid. Note well, only the potential is there. Given the history of the state PUC jurisdiction over electricity rates, and given a predisposition to contract federal authority, it is unlikely that the FERC together with the DOE will move to nationalize retail ratemaking, unless, of course, the dire predictions of regional blackouts are believed to be imminent.[101]

FERC ratemaking authority directly allocates abandonment costs, and there is a discernible trend away from protecting consumers toward splitting costs between ratepayers and shareholders, and toward apportioning costs among utilities participating in a nuclear project. The FERC is stepping in, to the extent that its formal jurisdictional authority allows, and is accommodating the various interests affected by nuclear abandonment. The FERC, like the majority of state PUCs, is trying to keep utilities financially afloat without imposing all of the costs on the ratepayers, particularly when the ratepayers receive no tangible benefits.

The FERC has adopted an accommodationist policy to cancellation expenses. In *New England Power Co.* the commission wrote:

> In the electric ratemaking area, the Commission's policy has been to allow public utilities to recover prudently incurred investment costs in a generating project that is ultimately abandoned or cancelled. . . . The Commission has refused to allow recovery of abandonment losses that were not shown to have been prudently incurred . . . . The majority of these cases, however, focused on how to amortize the losses incurred rather than on the prudence of the utility in incurring the losses.[102]

The allocation of cancellation costs is a delicate balancing of competing interests. It is also an example of rough justice and an imprecise weighing of conflicting political and economic variables. The decision maker who decides to accommodate the utility's owners and customers must grant to the utility enough money to stay out of bankruptcy and stay financially healthy so it can attract capital, and not enough money to promote imprudent overinvestment. The key to the calculation is a test of prudency. In *New England Power Co.*, the FERC adopted a liberal prudency standard, which says that the commission cannot substitute its judgment for that of the directors of the company unless management abuses its discretion.[103]

Pass-through of all cancellation costs is not automatic. The FERC has disallowed the unamortized portion of its investment loss in the rate base, although it can receive the investment back as expenses. Simply, investors are not allowed a rate of return under the theory that no plant that was used and useful resulted, but they recover expenses.[104] As the D.C. circuit noted:

> Because it would be inequitable to place on the utility the entire loss of expenditures prudent when made, FERC allowed recovery thereof over time. At the same time, FERC is charged with a duty to protect consumers against unreasonably high rates. Far from an inconsistency, FERC here struck a reasonable balance between the interests of investors and ratepayers.[105]

The FERC has recently ordered four operating subsidiaries of Middle South Utilities to share the $3.4 billion cost of Grand Gulf I in proportion to the amount of Middle South they currently sell. The effect of the decision is to apportion among various states the amount of the project that ratepayers must pay.[106] The impact of the decision is that for jurisdictional sales, the FERC has the power to apportion costs.[107]

The FERC, with the support of the DOE, also helps utilities with cash flow by allowing CWIP in the rate base. The FERC has adopted a rule that allows partial recovery of CWIP. All CWIP for pollution control facilities and fuel conversion facilities may be given rate base treatment and 50 percent of any other CWIP allocable to electric power sales for resale.[108] Similarly, the FERC allows utilities to recoup current costs incurred for permanent waste disposal where customers will not actually realize benefits for years.[109]

Originally, the FERC denied a utility's request to include costs for the permanent disposal of nuclear waste,[110] but it later reversed itself in response to the Nuclear Waste Management Policy Act of 1982,[111] which imposes determinate disposal costs on utilities. The FERC and the reviewing court reasoned that there is currently little likelihood that spent fuel will be reprocessed; therefore, permanent disposal costs should be amortized even though permanent disposal is years off.

3. Nuclear Regulatory Commission

The NRC has no formal authority over the allocation of cancellation costs. However, as the central policy-making bureau, its rules and regulations directly affect the costs of plant construction. The post-TMI period has been especially turbulent for the NRC—so much so that the agency's viability as presently constituted can be seriously questioned. Both the Kemeny[112] and Rogovin[113] commissions criticized the agency for its preoccupation with licensing and its inattentiveness to safety. Both recommended that the agency be reconstituted.

The NRC's immediate response to TMI was like everyone else's—the commission was alarmed and began to tighten safety requirements. As the case studies demonstrate, safety inspections were taken more seriously both by utilities and the NRC. Further, the NRC stepped up the number and amount of fines. Figures 5-1 and 5-2 graphically portray a noticeable increase in NRC enforcement actions in 1979, the year of TMI, and a significant decline around 1983.

The NRC continues to approve licenses. NRC enforcement actions have not acted as a deterrent as much as they have been an eye opener for utilities. Now, utilities must approach construction more cautiously. They must be particularly sensitive to quality controls and quality assurance during construction. And they must be equally sensitive to training and monitoring operating staffs.

Another costly post-TMI regulation was the NRC's adoption of emergency-preparedness rules.[114] Emergency planning can be seen as another layer of safety aimed at protecting persons on and near the reactor site. Prior to TMI, utilities were required to submit on-site plans. Just before TMI, a joint NRC-OPA task force released a report regarding off-site planning.[115] The need for such planning became apparant when Pennsylvania Governor Thornburgh advised pregnant women and preschool children to evacuate the area within a five-mile radius of TMI shortly after the accident.

The NRC and the Federal Emergency Management Agency issued a document providing guidelines for state and local governments and nuclear facilities for developing emergency response plans.[116] The NRC adopted final rules in the summer of 1980, under which no operating license will be issued unless a finding is made that there is reasonable assurance that adequate protective measures can and will be taken in the event of a

Figure 5-1. Number of Fines Assessed by NRC

years 1973 to 1984

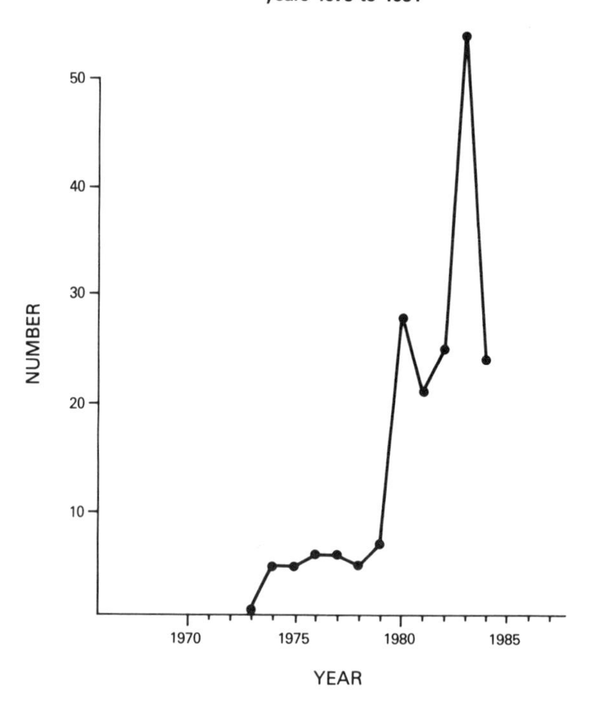

(Compiled from data contained in IE Enforcement Staff, Enforcement Actions: Significant Actions Resolved, NUREG: 0940 for years available, and other NRC data)

radiological emergency.[117] Emergency-preparedness requirements have been costly, particularly to Shoreham. The Shoreham plant began construction prior to TMI. Therefore, these rules were applied retroactively to it and, together with faulty emergency generators, are chiefly responsible for Shoreham's not being on line.

Emergency planning is a governmental quagmire. Not only are two federal agencies responsible for emergency coordination, but plans must be developed with state and local governments and with the owner-utility. With such a mix of jurisdictions, any participant has an effective temporary veto either by delaying or refusing to participate in emergency planning. When both the state of New York and the county of Suffolk refused to participate with Shoreham in establishing an emergency plan, LILCO developed its own plan by creating a literal shadow government. LILCO had to make provisions for evacuations, broadcasting, emergency health services, and myriad other governmental functions. Putting such a plan together is costly, and exactly how costs should be apportioned is not subject to a specific

Figure 5-2. Amount of Fines Assessed by NRC

years 1973 to 1984

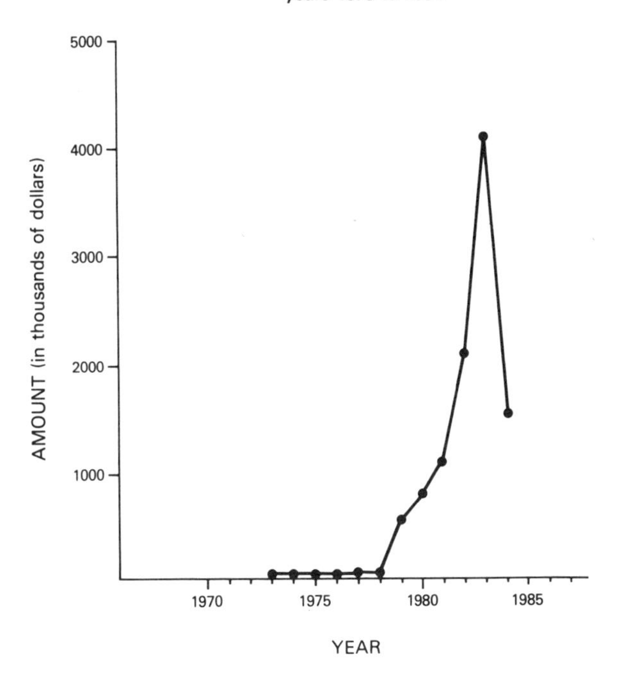

(Compiled from data contained in IE Enforcement Staff, Enforcement Actions: Significant Actions Resolved, NUREG: 0940 for years available, and other NRC data)

rule.[118] A General Accounting Office report asserts that most state, local, and utility officials agreed that off-site planning costs should be paid by the utility.[119] Most likely these costs will be passed through to the ratepayer.

The NRC's enthusiasm for emergency planning has waned. In two instances, the commission attempted to circumvent its own regulations and was checked by a federal court of appeals. In *Union of Concerned Scientists* v. *NRC*,[120] the NRC adopted a rule that the Atomic Safety Licensing Board need not consider the results of emergency-preparedness exercises in its licensing hearings. This rule had the effect of denying a hearing on emergency preparedness, which is, according to the commission's own regulations, material to issuance of a license. The reviewing court ordered the NRC to conduct a hearing consistent with the Atomic Energy Act. Also in *Guard* v. *NRC*,[121] the NRC sought to avoid following its own regulation regarding "arrangements . . . for medical services." The court held that the NRC's interpretation of the rule that the response plan merely list existing facilities was insufficient.

Since TMI, complacency has not been a watchword at the Nuclear

Regulatory Commission. Indeed, quite the opposite is true. The NRC is still caught in the position of encouraging the safe development of nuclear power while shedding its promotional image and continuing to license plants. Stepped-up safety inspections, increased fines and penalties, costly backfitting, emergency preparedness, waste disposal, and decommissioning regulations demonstrate the NRC's activist intervention into the construction and licensing processes. This intervention is costly, and, naturally, that has caused turmoil in the industry. Industry's instinctive reaction is to attempt to minimize regulatory costs. Such stepped-up activity has also brought internal turmoil to the agency. In response to critics who chided the NRC for its prolonged regulatory lag, the NRC undertook a major internal review by establishing the Regulatory Reform Task Force in 1981. The task force was specifically charged with:

—creating a more effective and efficient vehicle for raising and resolving legitimate public safety and environmental issues;

—developing means for more effective use of NRC resources in licensing new plants;

—avoiding regulatory uncertainty and placing unjustifiable economic burdens on utilities (and their ratepayers); and

—accomplishing the above without impairing protection of public health and safety.

The Task Force Draft Report, issued November 3, 1982,[122] reflects the pressures brought to bear on the NRC to tighten or streamline the regulatory process in an effort to reduce costs and step up licensing. The draft report was reviewed by an Ad Hoc Committee for Review of Nuclear Licensing Reform Proposals, a group of industry representatives, public interest groups, and state governments. Public comments were also received by several interested parties. Not surprisingly, the reactions to the draft report can be aligned neatly, with utilities and industry interests opposed to intervenors and individual interests. This positioning reveals the predicament in which the NRC finds itself. In an effort to follow its congressional mandate and in reaction to the change in public temperature, the NRC increased safety inspections and regulations. This increase has the direct effect of raising costs to the industry. To offset rising regulatory costs, the NRC, through the task force, also attempted to tighten regulations as a cost reduction measure. As a consequence, the task force suggestions lessen public participation in the regulatory process.

The task force report is divided into four major sections: Legislative Proposals; Administrative Proposals; Hearing Process; and Separation of Functions and Ex Parte Communications. Each section is further subdivided. The report and the comments it had engendered reveal the direction of NRC policy making.

a. Legislative Proposals

The task force recommended two major legislative changes, both of which would amend the Atomic Energy Act of 1954. Although the Nuclear Standardization Act of 1982 (Standardization Act) was drafted first, the Nuclear Licensing Reform Act of 1983 (Reform Act) was deemed to be of more immediate importance.

The Reform Act is intended to amend and improve the nuclear siting and licensing process by collapsing the present two-step (construction and operating) licensing process into one procedure. The construction license involves site and preliminary design approval. The operating license concerns final design. Both involve costly public hearings, and the overall lead time before a plant goes from drawing board to operational is twelve or more years. The one-step procedure contemplated by the Reform Act intends to make construction scheduling, design, and environmental review more certain and final prior to beginning construction. The one-step procedure together with early site approval and approval of standardized designs is a streamlining measure. The Reform Act also contains an important bow to federalism by including a section on certification of need for power by any authorized federal, regional, or state governmental organization.

While the ad hoc committee as a whole generally endorsed the Reform Act, there were two separate dissents. Most of the committee's comments were directed at the technical structure of the proposed reform legislation, and they advocated including backfit provisions to assure design stability. Separate views by a state official and a public-interest lawyer criticized the Reform Act for reducing the opportunities for public intervention.

The NRC also proposed the Nuclear Standardization Act of 1982 (Standardization Act). The proposal provides for design approval and stability of design for standardized nuclear plants, one-step licensing, and early site approval. The Standardization Act would add a new section to the Atomic Energy Act of 1954 empowering the NRC to facilitate early resolution of design-related issues. Although the NRC already has the authority to approve standard design, this new section provides explicit statutory support, encourages the development and use of standardized designs, and establishes standardization requirements. The intent of the act is that first approval is essentially final design approval for a design to be used at multiple sites with minimal adaptations. No one opposed design standardization. Before there is much to criticize, the NRC must adopt its own regulations. The standardization problem lacks immediacy, since no one predicts the need for new nuclear plants in the forseeable future.

b. Administrative Proposals

The most significant administrative proposal made by the task force concerned proposed backfitting regulations. On August 2, 1982 the task force published a proposed rule and policy on backfitting (SECY-82-326), which met with considerable industry opposition and was replaced by proposals contained in SECY-82-447. To industry, backfitting was a major cause of uncertainty and resulted in costly and questionable changes.

Defining backfitting is not unproblematic. The term can refer to requirements imposed either after a license is granted or after construction has occurred. Choosing a definition is relatively straightforward; more difficulty lies in determining what standards to apply. According to the task force, backfitting regulations should provide substantial additional protection of public health, safety, common defense, or security rather than mere marginal increases. The task force proposed a cost-benefit approach to backfitting regulations. Factors to be considered in the cost-benefit analysis include:

1) reduction in the risk of accident off-site releases;
2) consideration of the impact on occupational exposure;
3) costs of the backfit, including downtime and delay;
4) impact on safety due to changes in complexity and the relationship to other requirements;
5) resource burden on the NRC; and
6) difference in plant vintage type and design on the appropriateness of the backfit.[123]

The ad hoc committee unanimously agreed on the need for a disciplined, documented justification for the imposition of backfitting requirements together with a high level of review. The committee was not of one mind on the factors to be weighed in evaluating whether particular backfitting regulations were justified. The committee, with the exception of one dissenting member, understood that the six listed factors were to serve as guidelines. The dissenter interpreted the factors as limiting the use of the good judgment of the NRC staff. Separate views were stated by two ad hoc committee members. One, a state government representative, believed that the cost-benefit analysis, as proposed, did not adequately consider the public financial costs. The other separate view argued that the cost-benefit analysis provisions could not be adopted by rule. Rather, legislative action was required to shift the emphasis from public safety protection to cost balancing. Inherent in cost-benefit analysis is the propensity to quantify variables so that decision makers can make choices among competing plans. An economic bias confuses decision making because of the inherent inability to quantify all "costs" and "benefits." While cost-benefit analysis is a useful tool for gathering and highlighting data, its ability to function as a dispositive decision making rule, even for a subset of regulatory problems such as backfitting, may require congressional action.[124]

c. Hearing Process

The task force's approximately twenty-five hearing process changes were recommended in three parts: screening, hearing, and decision making. The screening and hearing changes involve the tightening of discipline for both participants and licensing boards. Changes in the decision-making process were directed at eliminating layers of review.

A major modification during the screening process was treating the

Atomic Safety and Licensing Board as a clearinghouse for all hearing requests, petitions to intervene, and proposed contentions. The report also proposed to upgrade the standing requirement and raise the threshold for the admission of contentions.

During the hearing, the task force recommended that the presiding officers be given more control over discovery, presenting evidence, and framing and raising contentions. It was proposed that the role of the staff be curtailed, and the staff, through a discovery request, could not be required to perform additional work beyond that needed to support their position.

The most significant suggested improvement in the decision-making process concerned the elimination of the Atomic Safety and Licensing Appeal Board as an independent, intermediate administrative appellate tribunal. Instead, the appeal board would be placed under the commission, where it would review licensing board decisions and recommend a course of action for the commission and would draft opinions for commission approval. The theory behind this transfer of functions was that important policy decisions that might arise during the course of the appellate review could be addressed directly by the commission.

The participation of intervenors, the report suggested, should be limited to those issues raised by them. Currently, intervenors may cross-examine, propose findings of fact and conclusions of law, and file appeals on any issue in the hearing.

The ad hoc committee favored raising the standing requirements for intervention to the judicial standard, supervising discovery, and raising the threshold for admissible contentions. Regarding the conduct of the hearing, the committee supported proposals to consolidate private parties, limit discovery of NRC staff, strengthen board supervision of discovery, and limit oral evidence and cross-examination, among other adjustments. Finally, regarding the decision-making process, the ad hoc committee supported the task force's limitations on participation by intervenors.

d. Separation of Functions and Ex Parte Communications

The task force also reviewed regulations pertaining to the role of the NRC staff in its relationship to the commission. Proposed rule changes were suggested in an attempt to relax the restrictions on commission-staff communications so that commissioners would have better access to the expertise of their staffs. Separation of functions is a doctrine that is designed to allow staff and decision maker complete autonomy to perform their work without influencing each other. Rules prohibiting ex parte communications fulfill the same function and are also intended to prevent off-the-record communications between the decision maker and outsiders as well as staff in adjudicatory proceedings.

Public comments on various aspects of the Task Force Draft Report were received from utilities, architects and engineers, trade associations, states, individuals, and intervenors. Predictably, utilities, manufacturers, trade associations, and architects and engineers were aligned, as were intervenors and states. Regarding the NRC hearing process proposals, understandably, the intervenor groups opposed proposals that had the effect or perception of curtailing public participation. Instead of narrowing issues, the intervenors favored expansion and more participation and proposed public funding of intervenor participation. The task force and the ad hoc committee proffered an NRC that fit squarely in the traditional model of bureaucracy.

CONCLUSION

State and federal regulation of nuclear power since TMI has been anything but uniform. Power and authority have been decentralized between federal and state governments. Within the federal government, there are signs that power is being recentralized. However, with the heightened sensitivity to safety issues and the promulgation of a series of safety-conscious rules, the recentralization effort has a long way to go. Further, within the states the problems surrounding plant cancellations are not being resolved uniformly. The lack of uniformity is characteristic of a developmental period.

The transition lacks uniformity because of the many competing dualities. The traditional approach no longer works, and, even if we can clearly see what the future looks like or should look like, the problems of the past remain to be solved. It can be reasonably argued that it would be unfair now to regulate utilities according to a postindustrial competitive market standard when those utilities invested in nuclear power when they were encouraged to do so by a regulatory system that rewarded capital investment. Likewise, it is unfair to impose risks on ratepayers who had no hand in making investment decisions. Accommodation, then, may become a temporary necessity but need not become a regulatory way of life. Otherwise, past theory will guide future action when the underlying facts do not conform theory and action. The transitional period is the time to critique past theory, understand the changes in present circumstances, and design future policy.

6. Responsibility and Liability

CANDIDATES FOR LIABILITY

The preceding chapters established the causes and consequences of the multi-billion-dollar failed decision to go nuclear. Again, the primary problem is: Who pays? In a perfectly fair world, those responsible for incurring costs without producing benefits should pay. In an imperfect world, correlating responsible actors with irresponsible conduct is not often easy, particularly when the state actively promotes the capital expansion of a complex high-technology industry. Frequently, an exact correspondence between conduct, cause, and consequence is lacking. The government's encouragement of the private commercialization of nuclear power, for example, is not necessarily precisely related to General Electric's design defects or to CG&E management failures. The imperfect correspondence between cause and effect and responsibility and liability is symptomatic of a bureacratic polity that is pulled in opposite directions.

As a capitalist democracy, we pay allegiance to laissez faire principles as long as the market functions relatively smoothly. Measuring "relative smoothness," however, is a political judgment, not a precise econometric equation. Government regulates when the market fails, and government regulation reflects both market principles and political judgments that may be at odds with each other. In the case of nuclear regulation, the market has collapsed because projects are now too costly. At the same time, the political response has been to decentralize decision making, thus raising costs. This conflict needs to be reconciled. The reconciliation of market needs and political desires must be based on a coherent theory of liability, spreading costs fairly and efficiently, and seeking contribution from active participants, not passive victims. The liability theory must respond to obvious and immediate economic interests. Also, the theory must consciously recognize the longer-run political and policy implications of a particular cost-spreading strategy.

Equity and efficiency, the respective sine qua non of politics and markets, are not always coterminous. Critics of modern bureaucracy claim that policy makers, in their search for efficiency, lose sight of fairness. Cost-benefit analysis, for example, is a much-used agency decision-making methodology which aggravates the conflict between equity and efficiency by favoring quantitative economic data over nonquantifiable normative variables. The tendency to trade off equity in favor of efficiency is more pronounced when a bureaucracy is charged with "economic" rather than "social" regulation. This distinction between economic and social regulation has the negative effect of subverting social and political equitable values. Nuclear regulation should neither concentrate on finances at the expense of more normative matters nor adopt a distinction that favors economy over equity. First, the history of nuclear regulation shows that it is motivated by political as well as by economic values. Second, the industry-government configuration should provide some measure of democratic participation, because the government should be representative of the common weal, not just government-industry interests. Third, the nature of a high-technology industry coupled with high-risk, low-probability dangers demands that decision making incorporate equitable and political values into the cost-allocation theory.[1]

Likely candidates for imposing legal liability can be identified by noting the actors in the nuclear drama. These include the government and its officials, industry and its personnel, and consumers and investors. The most likely targets for cost-spreading liability are ratepayers, the utilities' consumers. At present, utility ratepayers are forced to absorb most costs associated with nuclear plant cancellations through rate increases.[2] Also, the burden of abandonment expenses has been shouldered by a utility company's shareholders.[3] That ratepayers should be a targeted group is odd, considering that they have the least voice in the nuclear decision-making process and may receive no electricity for their payments. The oddity is explained by realizing that the strong force in nuclear regulation is build first, assess safety and financial risks later. In the early stages of nuclear regulation, the promotional policy had broad support, which has eroded because the country finds itself overinvested in nuclear plants whose safety is questionable and whose capacity may be unnecessary. The rationalization for imposing these costs on ratepayers stems from the belief that electricity is good, utilities are necessary, and therefore, electric utilities should be saved.

Liability has been imposed only lightly upon groups, governmental and

Figure 6-1. Responsibility and Liability in the Traditional Model

private, who directly participated in the initial planning, attempted implementation, and eventual demise of nuclear power projects. The next sections examine the roles of the participants in the abandonment-costs fiasco and the theories of legal liability available for imposing these costs. The traditional model of utility regulation depicts an unstable hierarchy of responsibility and liability, as described in figure 6-1. At the top of the hierarchy, primary responsibility can be attributed to "government" and "industry" for so heavily promoting the joint venture. The second tier of the hierarchy consists of organizations and individuals involved more directly with carrying out promotional policy decisions. This level includes: private firms such as reactor vendors, architects and engineers, and construction contractors; public utilities; and government officials and industry personnel. The last level comprises ratepayers and consumers, shareholders and bondholders, and taxpayers. These last groups have little say in choosing nuclear power, though they bear most of the burden. The hierarchy is unstable because liability is apportioned inversely to responsibility. Those most responsible, "government" and "industry," are least liable. Consequently, those least responsible, ratepayers, taxpayers, investors, and owners, are most exposed to financial risk.

FIRST TIER

"Government"

The regulation of the nuclear power industry is not a simple matter. The industry is complex, the subject matter is often technologically and scientifically uncertain, and the financial investment committed to the development of the industry is expended over a decade before any electric service is realized. In addition to the complexities inherent in the industry, there is an intricate overlay of government regulators. For now, we will rest content with using the intentionally ambiguous term *government* rather than more specifically defining one of several bureaucratic agencies. Chapters four and five described in more detail the role of specific agencies of the federal government and the role of state agencies in their attempt to regulate the commercial development of nuclear power. A point made previously must be reiterated: Government is the centripetal force holding the nuclear power industry together. Without government support, there would be no nuclear industry. Government thus should stand ready to take responsibility for its protective and promotional role. Whether it does or not remains an open and problematic question, because "government" is a generalized abstraction[4] behind which no independent, concrete entity stands. Nevertheless, a discussion of the abstraction uncovers certain layers of meaning about the relationship of law and modern government.

In connection with cost absorption or legal liability, *government* is actually a euphemism for taxpayers. Only in the most outlandish of situations, when a government official acts outside the scope of his or her authority, usually with malice or other equally gross conduct, will an individual be held financially liable. Therefore, the word *taxpayers* must be substituted for *government*. Occasionally, there is a distinction between national and state taxpayers. The reference will be to national taxpayers. In any event, taxpayers pay when government is liable.

"Industry"

The fundamental fallacy in the arguments favoring imposition of cost liability on government is that government as such does not exist. The word *government* is a surrogate conception for taxpayers. Similarly, *industry* as such does not exist. Rather, industry is a collective concept including various public utilities and private construction and manufacturing corporations. Just as government passes its liability on to taxpayers, industry passes its liability on to owners and investors—the firms' shareholders and bondholders—or to consumers of the firms' goods or services. Industry is no less responsible for the nuclear venture than government; therefore, liability should attach. However, industry personnel most responsible for participation in nuclear decisions are managers, who, like bureaucrats, are less likely to be stuck with abandonment costs. Managers are given a degree of immunity so that they can make decisions without worrying about liability. Instead, costs are passed on to investors and consumers, who have little real participation in firm policy. Furthermore, the financial insulation provided by government does not deter irresponsible conduct.

Imposing costs on industry may also appeal to an efficiency argument, but that is fallacious also. In a smooth market, costs are distributed to private-sector actors in the best position or most willing to absorb them. However, the nuclear market is highly artificial. Industry, in the activist state, is insulated from market risks by government support. The example of turnkey contracts demonstrates this principle. Reactor vendors took large short-term losses to create a long-term market. The gamble failed when the long-term market went bust. Presumably, the utilities should pay the cost overruns, but these are passed through by the ratemaking system. Similarly, the Price-Anderson Act encourages private-sector participation by providing a financial safety net. The act, which enjoyed broad political support in 1957, successfully brought money into the nuclear market. That political support has disintegrated, and the act may be altered severely when it comes up for renewal by imposing greater financial liability on firms. While it can be argued that it is both fair and efficient to impose increased liability on industry, because that is how markets ought to work, such a position ignores government's role in creating an enticing investment market. The efficiency argument also fails to recognize that the costs will be passed through to others rather than absorbed by industry actors.

THIRD-TIER JUSTIFICATIONS

There are two important and conflicting justifications for imposing costs on taxpayers, ratepayers, and shareholders. These arguments can be called third-tier justifications. The first justification is the argument from efficiency. Cost spreading to national taxpayers, or over third-tier actors, means that abandoment costs are spread as thinly as possible. This disbursement may seem desirable as well as fair. The rationale behind cost allocation to taxpayers is a recognition of the government's encouragement and promotion of the nuclear power industry as something done in the national interest. The government, as the representative and the embodiment of the public voice through democratic processes, has been actively engaged in the development of an industry that held great promise for our future. Nuclear technology was to be tamed by the transition from the destructive forces of its power to a clean, inexpensive, safe, and peaceful source of electricity. Concurrently with commercialization, the United States would be in the forefront of developing and controlling a technology capable of world destruction, thus maintaining superiority in the nuclear standoff. In this scenario, the nation's economy, energy supply, and national security could be vouchsafed. Our friend the atom had a friend in Washington. When the bright future did not pan out because of circumstances beyond anyone's control, the powerful ally in Washington would help rescue the needy industry. The unrealized benefits would be less painfully absorbed in small segments by the country's taxpayers. The taxpayers would feel little pinch and would be somewhat mollified knowing that they were participants in an unfortunately unsuccessful experiment in public policy. Similarly, cost allocation to shareholders and ratepayers is a recognition of industry's promotional role. Industry and its beneficiaries simply pay in proportion to the market risks they assumed.

The second justification is the argument from responsibility. Costs are distributed to taxpayers (or ratepayers or shareholders) because the government (or industry) accepts financial liability for its mistakes. This return to rugged individualism can be described as frontier ethics applied to modern technology or as a constitutive element of the activist state. Given the nature of regulated industries, government and industry should at least share costs, if the government does not pay the full tab. The government was instrumental in encouraging private-sector participation. Regulatory controls did not remain constant and, not infrequently, plant owners had to dump additional finances into a nuclear project to comply with the government's changing commands. Additionally, the legal regime institutionalizing government's nuclear policy reinforced the private sector's commitment of capital to nuclear projects. Rate regulation was an incentive to invest in large, capital-intensive projects, and a utility would not earn a return on investment until the plant was operational. The private sector, plant managers and operators, and shareholders and bondholders, had no real choice but to capitulate to the regulatory state's increasing demand for capital contribution. This scheme of government aid can be seen as an exercise of government responsibility or not so charitably described as a bailout. So painted, it is not only reasonable but fair for the government to stand ready to financially support the investment decisions it directed.

The arguments from efficiency and responsibility are rhetorical and undercut and reinforce each other. Choosing the responsibility rationale both supports and undermines the notion of efficiency. By accepting responsibility for losses incurred as a result of the failed decision to build nuclear plants, and by recouping those losses through taxes, the government spreads costs widely and, arguably, efficiently. Nevertheless, this scheme of taxation subsidizes the nuclear industry. Industry, then, is inefficiently spared the losses otherwise attributable to bad investment decisions. Likewise, if the efficiency argument is accepted, it supports and undercuts the argument from responsibility. Government accepts fiscal responsibility for its policies, but the subsidization of industry legitimizes and encourages irresponsible decisions to overinvest. The efficiency and responsibility justifications are also used to impose costs on shareholders and ratepayers. The more thinly spread the costs are, the less painful (more efficient) the payment. Also, industry and its beneficiaries thus accept responsibility for their promotional actions. The reason these two justifications can support and undermine each other simultaneously is that they are exercises in policy rhetoric that tell only half of the abandonment-costs story.

In no small sense is rhetoric a part of nuclear policy making. Rather, at this level of policy analysis, deep rhetoric often determines the policy outcome. Cost-allocation alternatives are best viewed as choices between different political and economic world views: a dialogue between politics and economics shapes policy choice. The market argument from efficiency and the political argument from responsibility conflict with each other, yet they are both caricatures of the same economically based story. Responsibility-based bailout is a picture of government enticement and dominance of a benign industry. The corresponding image of costs ever so thinly and efficiently spread is a portrayal of an unsuccessful, but heroic, public project. Both stories, however, have the same ending, with regulators saving the industry.

The dialogue between economics and politics yielding a policy choice of cost allocation to third-tier actors is too narrow. The essential weakness in the current dialogue about nuclear policy is that it has been dominated by economic arguments at the expense of political interests. Specifically, nuclear policy has not kept up with changes in a political climate that has grown skeptical of nuclear power. The failure of bureaucracy to respond effectively to changing conditions is the essential weakness in the law-follows-policy formula. When policy is institutionalized by law, dramatic changes in politics and markets work themselves slowly through the legal system, policy is fragmented, unfairness and inefficiency result, and the system experiences regulatory failure. All of these things are characteristics of a policy in transition. The political dimension of the nuclear policy and regulatory debate that honors democratic participation is slighted in an effort to resolve the financial predicament. The collective voice of third-tier actors, particularly consumers and taxpayers, is conspicuously missing from the responsibility and the efficiency justifications. Unfortunately, that voice is often too weak to affect policy.

The major obstacle to effective and constant public participation is the problem of collective action.[5] Individual citizens and shareholders are not readily compelled to organize and form coalitions to take on actors the size of the government and the nuclear industry. Public-interest organizations are too poorly financed, are too unorganized, and have too many divergent interests to compete against larger, more centralized interest groups.[6] Another obstacle is the semifalse belief in the objective nature of government oversight bureaus and of the technological establishment.[7]

Belief in the idea of scientific progress, where technology advances in a linear fashion objectively verifiable by experts, is comforting. In an effort to solve complex problems, the modern state facilitates the development of science and technology through its offices. With this set of beliefs and attitudes, complex decision making is taken away from everyday concerns and is given to experts. The promise of this type of bureaucratic state is that Congress delegates policy-making authority over selected issues to specialists. The bureaucracy then develops the expertise necessary to fashion correct solutions.

What is true about this description of an objective and scientific bureaucratic (or technocratic) state is that it is the premise on which agencies are created, *and* we, as a society, desire it to be true. Our desire is grounded in the hope that complex problems can be mastered by the experts. What is false about the picture is best exemplified by the subject matter of this book—the failure of nuclear power regulation. The vision of unassailable bureaucratic expertise is simply not true. The collective agencies of the government have not devised answers to what now appear to be intractable problems of nuclear power, such as plant licensing, decommissioning, and waste disposal. Misplaced public faith that the experts would provide answers has partially disenfranchised the citizenry from effective participation. The point emphasized here is that the imposition of cancellation costs on any group is a politically loaded policy choice. It is not a choice premised on a preordained vision of bureaucratic decision making, because that vision has failed.

The cost-allocation decisions studied here are expedient short-term solutions that do not readily respond to long-term problems caused by the close industry-government relationship behind commercialization. These short-term accommodations have the effect of continuing the motivation of industry and government to contribute capital into long-term projects because regulators provide a quick financial fix to buoy a sagging industry.

The weakness in the policy choice of imposing costs on government is that taxpayers are forced to take losses for which they receive no benefit and for which they had little or no input. The same weakness appears in the policy choice imposing costs on industry. These costs will ultimately be placed on shareholders and bondholders, or ratepayers and consumers, who enjoy little participation in policy making. Correcting this imbalance lies less in an outcome-determinative policy choice, for example, a choice between placing costs on ratepayers or shareholders, than it does in restructuring the decision-making system so that responsibility and liability are more closely aligned and more accurately reflect market and political interests. Consumers, investors, and taxpayers are the ultimate cost bearers. It is unfair to impose costs on them disproportionate to their voice in policy making, as the traditional model requires. Regulatory reform must seek a realignment of policy-making responsibility and financial liability.

SECOND TIER

The second tier has three qualities. First, because it includes indentifiable entities, the tier is not a generalized abstraction. Instead, specific organizations or individuals can be identified as the responsible actors whose conduct leads to recognizable consequences. Second, these actors are granted substantial immunity from liability, because they are carrying out orders of a larger entity. Individuals, such as government officials, are fulfilling the mandate of an agency created by Congress, and industry officials are carrying out corporate policy commands. Organizations rely on a similar claim. Private firms and public agencies both argue that they are pursuing the policy directives of popular government. Second-tier actors, the argument continues, are agents furthering the work of their principals; they are not acting on their own. Private individuals, most likely, will not pay for costs attributable to them directly from their pockets. Similarly, organizational immunity effectively exists, because liability is passed through to the third tier. Third, this grant of immunity is premised largely on myth. Myth has it that Congress carries out the will of the people, corporations carry out the will of their owners, and consumers have effective votes in the marketplace and at the polls.

Government and industry do not effectively coordinate the will of the electorate or the will of consumers and investors in their daily activities. Instead, government officials and industry personnel, in allegiance with organizational or institutional interests, exercise initiative and discretion independent of the desires of citizens and owners. The myth of organizational representation is the basis of the bureaucratic ideology immunizing government officials and industry personnel for acting on behalf of others. The myth also is the basis for the partial justifications of efficiency and responsibility already discussed. The reality of organizational independence exposes the myth. The undesirable consequence of the myth is the institutionalization of a policy choice allocating costs inefficiently and unfairly.

Federal Government

The Nuclear Regulatory Commission, as an example of a second-tier actor, is the primary federal agency involved with the licensing and development of nuclear power facilities. By authority of the Atomic Energy Act, the NRC is given responsibility to conduct inspections and investigations and to issue orders protecting public health and minimizing dangers to life and property. The NRC also has expansive powers over licensing, construction, and operation of nuclear power facilities for the protection of the health and safety of both plant employees and the general public. The NRC has adopted explicit guidelines and standards for inspection and oversight of the construction process. The agency's Office of Inspection and Enforcement develops the inspection policies and programs, carries out inspections and investigations to ascertain compliance, and enforces its findings through a variety of sanctions and penalties.[8]

The Office of Inspection and Enforcement has responsibility for reactor construction and the construction of nuclear power generation facilities. The regulations contain detailed technical provisions concerning plant construction, including provisions for an on-site resident inspector and staff, vendor inspection, and quality assurance and design control programs.[9] In practice, NRC personnel have been less vigilant than the aspirations contained in the regulations. The numerous accusations and investigations and the occasional admissions of misconduct during plant construction evidence the frailty of NRC oversight.[10] Misconduct aside, continuously changing regulations, licensing delays, and uncertainty regarding future standards for the construction of nuclear plants have contributed substantially to rising costs.

Safety inspections should be the very essence of the NRC's operations, though historically such has not been the case. Instead, the raison d'être of the AEC was to grant construction and operation licenses so that plants could be built and put on line. Since TMI, the NRC has exhibited a more vigorous role, although the attention now paid to safety and the promulgation of more stringent regulations is more a reaction to public opinion than it is a sua sponte change of heart by the NRC. The actions of NRC personnel during construction of new plants have often indicated that their interests are more closely aligned with the interests of the nuclear industry than with the public's interest in health and safety.

The NRC's safety inspection responsibilities carry three types of costs. Safety regulations themselves impose costs on industry that industry may not otherwise incur. In the language of economists, firms are forced to internalize externalities. Second, changing regulations impose extraordinary compliance costs. Third, any challenge to the regulations or any attempt to hold another entity liable for safety costs brings transaction costs such as litigation expenses.

Safety costs must be absorbed by someone. Consumers and investors are both interested in plant and reactor safety and in having NRC investigators perform their statutorily defined tasks. Both groups have a stake in uniform, predictable, reliable safety inspections, because these attributes lower costs. In the optimal situation, safety inspections are performed properly, and an acceptably safe plant results. The regulatory trade-off for a safe plant is to pass the safety costs to consumers.

Anything short of optimum safety raises costs. Changes in regulations raise compliance costs, and ineffective regulations result either in higher costs due to later regulatory alterations or in less safe plants. Who absorbs the extraordinary expenses, consumers or investors? Generally, consumers pay for extraordinary costs, because they are reflected in the price of the product or service. When the price of the product or service becomes too high and threatens the competitive position of the firm, then the investors absorb the loss or attempt to have the government pay for the conduct of their officials. Federal agencies and employees have some liability exposure for extraordinary costs resulting from regulatory changes or ineffective regulations.

In addition to extraordinary costs from changing or ineffective regulations, the NRC's safety inspection program may be deficient. Given the imprecision of NRC safety inspections, it is not uncommon for an inspection team either to miss violations and find them on a subsequent inspection, or to note the violations without taking any enforcement action. The case studies reveal that corrective action by utilities was not seriously considered until recently. Corrections mean delays, and delays mean higher costs. A plant can accept higher costs or pass them on to ratepayers or attempt to hold the government, manufacturers, or contractors liable. When a plant attempts to avoid costs, it incurs transaction costs. If a firm litigates and loses, for example, then it is forced to decide whether to raise the cost of products or services, pass the cost through, or reduce earnings. Even if the firm wins, the transaction costs must be allocated. In either event, inefficiency results, because the costs of products or services are increased, or profits are reduced when regulators fail to perform their duties. Inequity also results because the wrong class of people may pay. Consumers, ratepayers, and the firm are injured by the government's negligent inspections. The problems of changing or ineffective regulations and negligent inspections will be considered in a broader discussion of faulty safety regulation.

In the Zimmer and Marble Hill case studies, the NRC inspection teams cited both projects for numerous safety violations. However, the histories of both projects indicate a decisive change in attitude by the NRC regarding safety inspections. Prior to TMI, safety inspection reports were deathly mechanistic and apparently assumed little independent significance. One report on the Zimmer facility[11] reported "pop cans" and "apple cores" in the cable trays, hardly a significant find for a decade-long construction project. In another report, beer cans were spotted and noted with the same sense of urgency with which soda cans were mentioned. Yet, there was no discussion of drinking on the job. The magnitude of Zimmer's safety problems was noticed only after a whistleblower made enough noise. Projects have been granted licenses after thousands of violations.[12]

Many acts of neglect or improper performance of duties by NRC officers have been alleged. The conversion of Zimmer from a nuclear plant to a coal-burning facility is attributable in large part to the NRC's more vigorous response to allegations of previous shoddy NRC safety inspections. Specifically, an NRC investigator was formally alleged to have violated the law by:

> [M]ismanagement as defined in CFR 150.3(e); abuse of authority as defined in CFR 1250.3(f); perpetuating gross waste as defined in CFR 150.3(d); and perpetuating a substantial and specific danger to public health and safety.[13]

How responsible are NRC personnel or the agency itself for these various acts?

The principle of respondeat superior holds that employers are accountable for certain unintentional wrongs of their employees during the scope of their employment. However, the doctrine of governmental immunity prevents suits against the United States, as employer, for civil wrongs of its employees unless there has been a specific waiver of immunity. Sovereign immunity is based on the early common law notion that "the King can do no wrong." For government to function smoothly, effectively, and efficiently, immunity from suit is granted, and government will not be exposed to legal liability unless it consents. The Federal Torts Claims Act (FTCA), permitting litigation of tort claims against the United States in federal courts, is such a waiver. The act provides that the United States can be sued for the wrongful or negligent acts or omissions of federal employees, causing loss of property, personal injury, or death, "under circumstances where the United States, if a private person, would be liable to the claimant in accordance with the law of the place where the act or omission occured."[14]

The waiver of sovereign immunity is far from complete, and there are numerous exceptions to the act. One of the broadest is the "discretionary function" exception.[15] Protected discretionary functions or duties include "the initiation of programs and activities" as well as "determinations made by executives or administrators in establishing plans, specifications or schedules of operations." The exception is designed to protect federal agencies in their policy-making and decision-making capacities. Protection also extends to the acts of subordinates carrying out discretionary policies according to official directions.[16] The premise behind the exception is essentially protection of policy formation from interference and second-guessing by courts.

Acts by agency employees classified as involving "ministerial" functions, or nondiscretionary, operational duties do not fall within this exception under the FTCA, and liability rests with government. Judicial decisions indicate that courts have failed to reach a consensus as to what is, or is not, discretionary in nature. While some courts have given *discretionary* a very broad interpretation, others hold that the term should be restricted to "non-routine, essential judgmental, policy-type decisions."[17]

Two general types of allegations for faulty safety regulations can be made. The first complaint is that inspectors failed to perform the job assigned. Allegations that inspectors negligently performed readily identifiable tasks would fall into this category. The second category of allegation is that the NRC failed to design a proper safety inspection system. The difference between these two sets of complaints is liability. Under FTCA cases, the NRC would be liable if one of its ministers improperly completes his or her checklist. Generally, they are not liable if their system of inspection is faulty, a deeper generic problem. Naturally, safety inspections and accompanying problems are not so neatly divided between ministerial and policy-making functions. Discretion, and the discretionary function exception, occupy this gray area between the two poles. Before the federal government is liable for the negligent acts of NRC personnel, the acts complained of, such as improper promulgation or enforcement of safety standards, must not be based upon the exercise or performance of a discretionary function.[18] The NRC engages in both discretionary and ministerial activities. The agency's Office of Inspection and Enforcement develops policy and makes decisions concerning the regulation of new plant construction. Under the discretionary function exception, NRC officers or directors would be protected from liability in their policy-making functions, such as determining the safety standards required for new plants, deciding the types of inspections performed on nuclear power plants under construction, and supervising inspections in the field.

In a precedential decision, the United States Supreme Court significantly circumscribed the liability of federal officials under the discretionary function exception by indicating that no liability can be imposed because the policy itself or the method chosen for its implementation was faulty. In *Dalehite* v. *United States*,[19] the plaintiffs sought to recover damages for deaths resulting from an explosion of ammonium nitrate fertilizer produced under a program established and controlled by the United States government. The Court held that since the decisions to produce the fertilizer and to govern how it should be produced were made in the exercise of judgment at a planning rather than an operational level, they fell within the discretionary exception. This interpretation means that NRC officers would not be liable, for example, for failing to require certain tests or inspections as part of overall design controls or quality assurance programs for all nuclear plants constructed. Recently, the Supreme Court adhered to this protective standard in a case exonerating government officials from liability for improper certification of a type of aircraft for aviation.[20] NRC personnel would not be protected by this exception, however, when their actions indicate a failure to follow the prescribed agency policy and regulations such as improperly performing a required inspection.

On-site plant inspections, as opposed to policy making in NRC headquarters, would be the most fertile source of government liability. The courts, however, have not developed a uniform standard. In *Blessing* v. *United States*, injured employees sued the United States under the FTCA, alleging that OSHA inspectors had negligently inspected the employer's premises in dereliction of the standards imposed by the Occupational Safety and Health Act. The court stated, "When standards to be applied involve objective and known criteria and require only exercise of scientific and professional judgments that are reviewable under tort law, Government is not shielded from tort liability by the discretionary function exception."[21]

The NRC inspection and investigation programs detailed in the regulations are comprehensive, leaving little discretion to the individual inspector. The fault, however, can be either in the inspection or in the process. When NRC officers are conducting specific inspections and assessing quality control programs during the construction of new nuclear facilities, the agency is vulnerable for negligently executing nondiscretionary "ministerial" duties, but the NRC is not liable when engaged in protected judgmental activities.

The Federal Torts Claims Act also requires that the act or omission by the government employee be tortious under the law of the state in which it occurred. Most federal statutes and regulations, including those under which the NRC operates, create legal duties that state law does not impose upon private citizens; therefore, they cannot trigger liability under the FTCA. Commentators have argued that this requirement should be repealed because of the increasing number of federal laws lacking counterparts in the private tort law of the states.[22]

The law of the state in which the canceled nuclear plant is located governs questions of proximate cause and negligence. In plant cancellation cases, there are a number of intervening factors, such as delays and poor-quality work by contractors, technological and design errors by architects and engineers, and poor judgments by project managers, contributing to the cost build-up of later-canceled plants. The concurrent actions and omissions of these parties complicate the task of isolating those costs attributable to NRC nonfeasance or misfeasance. The cases indicate that intervening and superseding negligence of other parties may absolve the United States liability under the FTCA.[23] To impose liability upon the United States for improper inspections by NRC personnel, the court must examine the surrounding circumstances in each instance and determine what portions of the abandonment costs were proximately caused by the officer's negligence. The United States can be held liable for that portion of the costs associated with NRC negligence in performing the operational, nondiscretionary inspections and supervisory activities in accordance with the principle of aligning liability with responsibility. If the NRC was at fault, and the consequence is higher costs, then the government should contribute its proportionate share.

A serious legal question exists as to whom the duty to inspect runs. Does the NRC owe a duty to ratepayers for higher rates due to faulty inspection or to the utility for increased costs? The government agency must owe a legal duty to the claimant before asserting liability against the agency. To date, one case has been brought against the United States under the FTCA for negligent NRC inspections. TMI's owners, General Public Utilities, alleged that the NRC had a duty to warn of design hazards.[24] GPU argued that prior to TMI a similar accident had occurred at the Davis-Besse plant in Ohio and that the NRC's failure to warn was negligent. Further, GPU asserted that the NRC had approved TMI's construction plans, which contained the same PORV value at fault at Davis-Besse and later at TMI. The district court rejected the government's assertion that it owed no duty to GPU, and held rather that its duty inured to the public:

> An evaluation of the relevant statutes, regulations and legislative history established, however, that the relationship between the NRC and the nuclear industry contains elements of symbiosis, which *may* form the predicate for imposing liability.[25]

The district court avoided deciding whether the failure of a duty to warn or the issuance of a license for a plant with a faulty valve fell within the discretionary function exception by certifying the issue for interlocutory appeal. The court of appeals held that the utility's claim was precluded by the discretionary function exception.[26] A number of cases, dealing with a variety of federal agencies, have found that a duty to inspect exists, but not all have imposed liability for negligent inspection or failure to inspect.[27]

Other legal theories may not be available for placing costs of plant cancellation with the federal government. Because of the hazardous nature and additional safety and health concerns surrounding the use of nuclear energy, it could be argued that the NRC should be held to a higher standard when dealing with this potentially dangerous process. Strict liability however, has been proscribed by the Federal Torts Claims Act and the *Dalehite* decision:

> Since the Act may be invoked only on a negligent or wrongful act or omission of an employee, it created no absolute liability of the Government by virtue of its ownership of an inherently dangerous commodity or property, or of its engaging in an extra-hazardous activity.[28]

Another exception to the FTCA precludes government liability for claims arising out of misrepresentations.[29] This exception prevents federal liability for alleged NRC misrepresentations to the public, utility owners, utility commissions, and others connected with the building of a nuclear facility.[30] The FTCA also eliminates government liability for the intentional torts or crimes of its employees.

In rare instances, liability may be imposed upon negligent NRC officers in their individual capacities for some costs associated with plant cancellation. Government employees are clothed with varying degrees of immunity, depending on the type of function being performed and the source of law allegedly violated. Absolute immunity has been conferred on "judicial," "legislative," or "prosecutorial" acts. Qualified immunity has been available to bureaucrats who engage in quasi-judicial, quasi-legislative, or quasi-prosecutorial acts.[31] In *Scheuer* v. *Rhodes*,[32] the court discussed the level of immunity granted to a governor and indicated that more immunity will be provided for those higher-level officials involved in discretionary activities. The court stated that in varying degrees, a qualified immunity is available for officials, "the variation being dependent upon the scope of discretion and responsibilities of the office and all the circumstances as they reasonably appeared at the time of the action on which liability is sought to be based."[33] NRC inspectors would be given a more qualified immunity when they perform at a low, operational level that involves simply following the mandates of the regulations, and more immunity the larger their policy-making role.

State Government

State public utility commissions deal directly with nuclear abandonment costs. PUCs concentrate on the back end of the nuclear construction cycle by deciding how much a utility can charge its customers. Through the rate structure, a utility earns revenue for operation, maintenance, and additional projects. Unlike the federal involvement with radiological matters, the primary functions reserved to state public utility commissions are rate regulation and general supervision of nonsafety matters. Only rarely do PUCs directly involve themselves with construction or abandonment decisions. Utility managers have the initial responsibility to decide to build or abandon a plant. Once the managerial decision to initiate construction is made, the federal government does necessary licensing at the front end and supervision during construction and operation.[34] After the plant is built and on line, the state steps in to ensure that utility ratepayers are charged fair and reasonable rates for energy used.

Because of the bifurcated regulatory responsibility shared between state and federal governments, a state's position is somewhat attenuated. States follow the federal lead as the management decision to construct a nuclear plant is given approval by federal licensing authorities. Nonetheless, a state's role cannot be minimized. During a rate hearing, PUCs are obliged to gather facts, consider evidence, and, based on the record before them, establish fair and reasonable rates to be charged by a utility. PUCs have the power to determine what factors or items should be included in a utility's operating expenses and rate base and what items are not recoverable. This discretion allows the commission to exclude costs of utility property not to be included as a basis for the utility's revenue requirements. More important, state commissions have the power to exclude from the rate base costs accrued as a result of imprudent management. Management prudency is the central issue of rate regulation regarding canceled projects.

In the exercise of their ratemaking power, utility commissions are given broad discretion, but they must act in a lawful and reasonable manner and in substantial conformity with controlling constitutional and statutory provisions. The commissions act as impartial and independent tribunals. Should costs associated with canceled nuclear facilities be unjustly included by the commission in the rate base, the remedy for utility customers is a court appeal. Understandably, PUCs have vast discretion given the complex nature of ratemaking methodology and conflicting policy arguments in favor of and against including costs of canceled plants in the rate base.

States can be made liable for the wrongdoing of state officials or agencies in much the same way as the federal government is liable for acts of its employees. The doctrine of sovereign immunity applies to the states, as well, permitting no state liability in tort unless consent is given. Because of federal preemption of construction and the discretionary nature of PUC activities in determining rates, placing liability on state public utility commissions for costs of plant abandonment has not been attempted.

The reason for insulating state agencies from immunity in the area of rates and ratemaking is even more compelling than the rationale for the discretionary function exception. First, state public utility commissioners are engaged in a quasi-judicial function. The commissioners preside at hearings where different sides present evidence about competing policies. Even though state utility commission hearings are notoriously proindustry, it cannot be said that this bias taints commissioners, whose duty is to apply laws reflecting that bias. Even commissioners who may be sympathetic to consumer interests must confront arguments that advocate either saving a utility or balancing costs. Official decision makers must have freedom to exercise their judgment without fear of sanction.

The individual members of PUCs are state officials and are accountable for wrongs committed in office. Those wrongs, however, must be outrageous or criminal before liability attaches. A commissioner, for example, may not be involved in any business activity preventing him or her from being fair and impartial in carrying out his or her duties. The usual remedy for this type of misconduct is disqualification, not a monetary award. A federal statute is the primary vehicle for imposition of liability upon individual state employees for misconduct in carrying out their official duties.[35] Under this statute, a commissioner must knowingly or recklessly deny constitutional rights. The statute cannot be used to correct errors of policy judgment. Ratemaking involves more than just determining fair dollar amounts. Rate decisions require high levels of policy analysis about how costs for abandoned nuclear plants should be allocated. The commission's duties are quasi-judicial and quasi-legislative and are therefore protected to a greater degree than those of government officers executing operational duties. Provided the utility commissioners act in good faith in executing their duties, official actions receive a high level of immunity, insulating commissioners from personal liability.

Suing government for abandonment costs is an attractive alternative if we realize the magnitude of the role that government has played in pushing a pronuclear policy. We may even go further and say that government's involvement was irresponsible given the magnitude of the mistaken policy. Such assertions are simply not enough to hold government officials personally liable for what is fundamentally a mistaken policy choice. Government and its officials are given leeway to make policy mistakes. Before officials are held liable in their individual capacity, they must exceed the scope of their employment and must step outside their protective mantle of immunity. Even if government and its agencies, as opposed to individuals, were held liable, the victory might well be Pyrrhic, since taxpayers pay in the end. The fallback position is that spreading costs to taxpayers is the most efficient way of cost allocation. Although taxpayer allocation may be the least painful, it is not the most efficient, because industry escapes liability, and a bailout provides little incentive to lower costs. The next group facing financial exposure consists of second-tier industrial actors.

Firms

A nuclear power plant is designed by architects and engineers, built by construction contractors, and supplied by reactor vendors; these persons, together with the utilities, are the industry's organizational actors.[36] To the extent that any of their work is defective, the repair or replacement of the defects adds to the bargained-for cost. Likewise, delays in construction due to supply interruptions or regulatory lag also increase costs. The rules for imposing costs over and above those willingly and voluntarily agreed to by contracting parties are determined by tort and contract principles. During the turnkey period, for example, the contractors and vendors contractually assumed the risk of overruns. After that time, the risk shifted to the utilities, again by contract. Prior to the heightened sensitivity to costs after TMI, most cost overruns were passed through to ratepayers. Now these pass-throughs are too expensive to be done automatically. Automatic pass-throughs are even less justifiable when no service results.

Although the nuclear power industry has been actively building plants for nearly three decades, construction litigation was rare until recently.[37] Even today, litigation among members of the nuclear industrial community is a new and developing field of law. Nuclear litigation demonstrates that the law-follows-policy formula has pervasive effects throughout the legal system. Nuclear regulation, generally considered to involve public law, has a direct impact on how private law rules are interpreted and applied. New litigation theories and strategies are emerging peculiar to nuclear power because of the size and length of time of construction projects and the uneven distribution of experience among the relatively few participants. Allocating contract and tort liability is further complicated by the role of government in encouraging construction and by regulator negligence. Nuclear construction litigation, then, involves opening up the contracting process, reassigning market risks, and attempting to impose tort liability in light of the public nature of nuclear power.

Although utilities and consumers may be willing to sue other industrial actors, these suits are procedurally and legally difficult and very expensive. Design defects and changes, construction delays and stoppages, and faulty workmanship and materials have a tremendous impact on the cost of nuclear plant development and, consequently, on nuclear plant cancellations. Many of these problems are accompanied by regulatory foul-ups or changes in safety or technical requirements. Other problems may be due to the newness and ever-changing state of nuclear technology, such as a reactor vendor's rushing to put its product on the market without proper testing,[38] or simply dissatisfaction with product or performance.[39] The concurrent and interrelated nature of the acts and omissions of the parties at the nuclear plant site creates difficulties in sorting out who is responsible for what costs. The question is whether conventional theories of construction litigation will govern disputes in nuclear plant construction, forcing industry actors to bear portions of the abandonment costs attributable to them. Much of the litigation between utilities and corporations with whom they have contracted is an exercise in post hoc characterization. The utilities' posture is that they relied to their detriment on the expertise of more experienced industry personnel.[40] The utilities did not have technical and scientific sophistication in nuclear power and were put in a dependent position. The utilities' argument is that they are victims of the long collaborative and self-insulating relationship between government and industry. Therefore, they are entitled to compensation over and above that provided for in contracts.

The response by contractors and manufacturers is more staightforward and traditional. The contractors, vendors, and architects and engineers argue that the right, duties, and liabilities of the parties are set out clearly in contracts voluntarily negotiated by free large commercial enterprises.[41] Liability thus has been predetermined at the negotiating table and should not be rehashed in a courtroom.

Principal litigants at this tier are plants and contractors. Litigation will attempt to shift costs to particular firms, as opposed to imposing personal liability on individuals. Firms saddled with costs either incorporate transaction costs in their product or reduce earnings to shareholders. The principal cost bearers are, therefore, consumers and investors. If overruns associated with abandonment costs fall within normal business risks, then ratepayers have few complaints. The fact is that cost overruns attributable to abandonment costs are ten times estimated cost projections for plants, and industrial actors are allocating costs inter sese. The allocation of these extraordinary expenses must take into account the nature of the collaborative government-industry relationship.

A liability theory maximizing efficiency and fairness will channel costs to responsible parties instead of passing them down the line. Consumers and investors argue that poor workmanship and design cause higher costs. Shareholders can sue a utility's managers on behalf of themselves as owners.[42] Ratepayers and consumers are in a more difficult position, because they are a step removed from the contracting process. They must assert a direct tortious injury, allege a statutory cause of action,[43] or claim that they are intended third-party beneficiaries of the contract between owners and builders. So far these cases have met with little success.

In *Pennsylvania* v. *General Public Utilities Corp.*,[44] state and local municipalities sought recovery against owners, operators, designers, and builders of the Three Mile Island reactor in connection with the accident at the facility. The court stated that private litigants may not maintain an action to enforce the provisions of the Atomic Energy Act but may assert state tort causes of action against the utility. The plaintiffs alleged negligence and willful misconduct in designing, constructing, and operating the nuclear facility and sought monetary damages for expenses incurred as a result of the accident. Consistent with the safety/financial distinction of *Pacific Gas & Electric,* the court of appeals overruled a previous summary judgment,[45] remanded the case for further development of the record, and said:

> This case presents important and at least factually unique issues involving the potential libility of designers, builders, owners and operators of privately owned nuclear energy electric generating plants to pay damages to state and local governmental units and agencies in the event of a statutorily defined nuclear incident. To pass upon these issues requires a record more complete than that before us and the district court.[46]

The case illustrates the difficult legal position in which third-tier actors find themselves. First, liability theories for nuclear abandonment costs are novel. Because they are undeveloped, courts are unfamiliar with the theories, and a normal judicial response is to dismiss the action. Second, the status of the third-party claimants, for example, consumers, is unclear. Should they have a legal right to sue contractors when the real injury, either due to breach of contract or tort, lies with the firm? Third, it is difficult to find the appropriate forum. Should consumers file suit in federal or state court, or before a federal or state agency? Thus far no forum has been particularly friendly.[47]

County of Suffolk v. *Long Island Lighting Co*.[48] is the case currently most on point for nuclear abandonment-cost issues. The county, on behalf of itself and 800,000 other ratepayers of Long Island Lighting Company, brought suit against the designers, suppliers, contractors, and owners of the plant. The plaintiffs claimed that design and construction of the reactor were defective and asserted causes of action for negligence, breach of contract and warranty, and misrepresentation and concealment. The court noted that the only damages sustained by the ratepayers were increased rates reflecting higher construction costs for the Shoreham facility and that the appropriate forum was the PSC. The relief sought included independent safety and health inspections and an injunction prohibiting the utility from charging rates including construction costs attributable to design or construction defects. Although the plaintiffs claimed that the suit was "pocketbook" action for economic damages, the court found that the plaintiffs were actually motivated by safety concerns and were seeking indirectly to regulate the radiological safety of the plant. Both the district court and the court of appeals agreed that the ratepayers' safety-related claims were preempted by federal law.[49]

In addressing the economic issues relating to plant cost, the court held that it could not entertain plaintiffs' claim for rate relief because the public service commission has exclusive jurisdiction over all requests for retrospective and prospective rate relief. The common-law causes of action were precluded because state law did not allow recovery for economic loss. Breach of warranty and contract claims were unsuccessful because the utility ratepayers were not in privity of contract with the defendants furnishing the faulty work. LILCO's objective in entering into contracts with suppliers and contractors, reasoned the court, was to benefit its shareholders upon completion of the facility, not to benefit the utility's customers; therefore, suit was precluded.

Third parties seeking to recover against contractors and designers for incurred abandonment costs will confront many of the same issues raised in *County of Suffolk v. LILCO*. Many of the common-law causes of action will be frustrated by the lack of privity between ratepayers and contractors. The rate relief avenue is frustrated because of forum unfriendliness and complexity of the issues that are heard in PSCs. Relief before the NRC is precluded by law. Ratepayers occupy an odd status. They are involved in neither the policy-making nor contracting stages but are all too likely candidates for liability. Shareholders occupy a similar status. They are little involved with planning or operational decisions and may very well pick up the tab. In an effort to avoid abandonment costs, either shareholders can sue in a *County of Suffolk* type suit, with rather dim prospects for success, or they can charge management with a breach of fiduciary duty.

Managers

The initial decision to build a nuclear plant, as well as the decision to continue construction or later cancel or convert, rests with utility managers. Utility managers assess the need for power, as well as choice of architects, contractors, and vendors. Corporate managers are responsible for overseeing construction and manufacturing processes. Management has a duty to avoid unnecessary costs. The duty runs both to investors and to consumers.

Managers' duty of care is a channeling device. When the managers satisfy their duty, the costs of their decisions are passed through. The short term interest of consumers is to prevent this pass-through. The interest is short-term because if the firm is forced to absorb too many costs, its corporate existence is threatened. If a utility is forced into bankruptcy, chances are slim that electric service will stop. Chances are much greater that the cost of the service will rise in an effort to strengthen the old utility or establish a new one. Corporate bankruptcy can cause a similar dislocation, depending on the nature of the market for the firms' products. The removal of a major partner in an oligopoly will raise costs. Shareholders, in contrast, try to keep costs in the product to avoid suffering losses. In either case managers pass the costs on to someone else.

The pass-through mechanism is being questioned, and suits charging management misconduct are pending.[50] These are not easy suits for ratepayers or shareholders to win. The rate structure is tilted in favor of allowing, indeed encouraging, management to invest in risky capital expenditures. The rate structure, with its capital investment incentive, is significantly responsible for the poor state of the electric industry.[51] With the state's imprimatur on management's investment decisions, it is difficult to prove misconduct. By pouring money into nuclear plant construction, management is doing exactly what it is required to do by law. Utility management, because of a utility's quasi-public status, has de facto immunity much like that enjoyed by government officials. This protection does not mean that managers can run a utility into the ground or invest in unneeded facilities.

The burden of paying for unused facilities or cost overruns has been placed on ratepayers, so that shareholders have not been injured and therefore do not have a claim against utility management. However, with the advent of cases such as *Office of Consumers' Counsel*[52] where the utility was not allowed to pass construction costs on to ratepayers, shareholders are directly injured. Shareholders suffer because the utility uses profits to pay construction costs of the abandoned plant instead of paying dividends to shareholders. Shareholders can claim that corporate management is liable for failing to meet their duty of care in making the decision to build a nuclear power plant, for not abandoning the construction of the plant in a timely fashion, or for abandoning the plant when it might have been wiser to complete it or convert it to a coal-using plant.

Private corporate management owes a fiduciary duty to the corporation, and thus to the shareholders. The fiduciary duty is one by which corporate management must act in the best interests of the corporation, meaning that management has an obligation not to misuse, waste, or mismanage the capital and property of the corporation. Fiduciary duty directly relates to abandoned nuclear power plants and to the issue of whether the decision to build was a waste of corporate funds. The burden of proving mismanagement and breach of fiduciary duty is on the shareholders, who must overcome the presumption that management made the correct business decision, acting in good faith and honest judgment.

Public utility managers enjoy an added degree of protection. Because utilities are regulated, managers are more circumscribed than their private-sector counterparts and must follow rules and regulations established by regulatory authorities. In exchange, they receive the added protection of being insulated from liability, unless their management decisions are imprudent.[53] Utility management will not be found liable for good-faith mistakes in business decisions and judgments. Historically, the prudent-investment test developed as a way to calculate the revenue requirement for the utility.[54] It is a problematic test for imposing liability on shareholders.

Presently, there are no cases in which individual managers were held liable for an abandoned plant's cost. A PUC or a court first assesses whether utility management made a prudent decision to build the nuclear plant and a prudent decision to abandon the project. Decision makers usually find that management's decision to build a plant was prudent and reasonable given the data available in the early 1970s and that the decision to abandon the plant was reasonable in light of the decline in demand for electricity and because of regulatory uncertainties later in the decade. Reliance on government or industry data is a bit surreal insofar as the data are self-fulfilling. Government and industry want data to justify investment decisions. Whether one uses data from either source, the message is the same, and the reliability of the data can be questioned. The problem is that these sources control the flow of information sustaining management decisions and protecting them from liability.[55]

THIRD TIER

The third tier of liability for abandonment costs comprises those persons most likely to pay for the nuclear mistake. Taxpayers have already been mentioned, and they are liable whenever the government is held responsible. There are two remaining groups—consumers and investors. In the second tier, the primary industrial actors were involved with intramural fighting over fault. To the extent that a public utility, the entity responsible for the decision to build or abandon a plant, is unable to recoup its losses from other sources, the costs are passed on to ratepayers or are suffered by utility investors. To the extent that private firms are stuck with abandonment costs, the expenses will be visited upon shareholders or bondholders or upon consumers of the firms' products or services. Therefore, the single most important confrontation is between consumers and investors.

Consumers versus Investors

Consumers and investors share a common fate. These groups are last on the list in terms of responsibility and generally lack an opportunity to choose policy or direct corporate strategy. Still, they are the most likely candidates upon whom the abandonment-cost burden will be imposed. Cost-allocation discussions usually pit these two groups against each other, moving debate away from the primary decision makers and into a liability analysis dissociated from any notion of responsibility. Corporate policies with lessened liability are inefficient. Imposition of liability without correlative responsibility is unfair.

The case for imposing costs on ratepayers and consumers is similar to the argument made for cost spreading to taxpayers. Ratepayers can more easily absorb costs. Further, ratepayers are beneficiaries of keeping utilities strong and viable. After all, ratepayers must have some assurance of reliable service. Dependable, continuous service avoids problems accompanying brown-outs or black-outs, and, more important, utilities are required by law to maintain service. From the utility's point of view, they are doing only what they are required to do by law—provide electricity. Proposed nuclear facilities are part of a long-range plan that was not successful because of unpredictable national and international market forces. The price hikes of the 1970s, together with conservation measures, resulting in declining demand have made nuclear plants no longer necessary. These arguments constitute the case for placing costs on ratepayers but are not universally accepted. Some states as a matter of law refuse to place these costs on ratepayers under the theory that ratepayers should pay only for the electricity they receive.[56] Other states refuse as a matter of policy. The hard question that must be addressed is whether the theory of cost-of-service pricing, where the ratepayer pays for what is received, should be retained by a regulatory system that must supervise an industry with lead times of over a decade and with evolving technologies.

Consumers are similarly situated. If the price of a firm's goods or services is too high, then the consumer can choose an alternative, to the extent that alternatives are available, or consume less. Here lies the difference between consumers and ratepayers. The ratepayer's freedom of choice is constricted. While it is true that ratepayers can use less electricity, their elasticity of demand may not be as great as that of a consumer of a private firm's construction services, for example.

The case for imposing liability on shareholders and bondholders is based on hard pragmatism and faulty logic. Most electric utilities, like other private corporations, are investor-owned. Shareholders are the owners and theoretically make decisions that run the corporation. Shareholders, therefore, are responsible for corporate decisions. The fallacy in the syllogism is that the middle proposition is not true for large modern corporations. Shareholders technically own the utility, but they do not run it. Their status as owners has significance only vis-à-vis corporate finance. Shareholders are more accurately described as investors basing investment decisions on corporate pronouncements concerning the health of the utility. The daily as well as long-term operations are handled by utility managers. Bondholders are similarly situated, because their investment decisions are based on information controlled by government and industry. A choice between cost spreading to ratepayers or investors is grounded in a relative policy choice designed to minimize the burden of liability. Imposition on ratepayers spreads costs more thinly and saves widows and orphans holding utility stocks and bonds. Essentially, determining who will bear the abandonment costs is a balancing of the competing interests of investors and ratepayers. These two groups are the "victims" of a variety of situations ranging from delays in construction to mismanagement to negligent government supervision for the purpose of maintaining the financial integrity of the utility.[57]

THE TRADITIONAL MODEL REVISITED

Note the zero-sum nature of the abandonment-costs game. The billions of dollars attributed to overinvestment in nuclear plants must rest someplace. According to the three-tiered structure of liability described here, the most likely resting place is at the bottom of the liability ladder. *Government* and *industry* are ambiguous references to two collectivities lacking precise identities. Both groups effectively avoid imposition of liability costs. The government easily passes the costs through to taxpayers. The nuclear "industry," as such, does not have a fine line of demarcation. Contractors, vendors, architects and engineers, and utilities use two powerful arguments for avoiding payment. First, industry argues that the fault lies with government's pushing a policy too hastily. Second, utilities assert that the law requires them to invest and allows them to earn a return on their investment from ratepayers. The inequity in this scheme of liability lies in an absence of responsibility from decision making. Those with the least participation, the persons in the third tier, become the targets for the abandonment costs. A liability theory honoring the virtues of responsibility and participation must be central to any regulatory reform. More concretely, neither consumers nor investors should be forced to pay for mistakes of firms' managers. Likewise, taxpayers should not be scapegoats for government errors.

The lesson of this chapter is that there is an inverse ratio between liability and responsibility endemic to nuclear regulation. The reasons marshaled in favor of the traditional model are based on abstractions and myth as much as they are on pragmatic justifications. Someone has to pay abandonment costs. After "government" and "industry" are revealed to be conceptualizations, and when it is shown that organizations and agents of government and industry are well immunized against liability, then the real fight about cost allocation is waged among ratepayers, consumers, shareholders, bondholders, and taxpayers. Imposing abandonment-cost liability on this last tier of responsibility may be unfair, but it is efficient, so the argument goes, because there is no one else left. The best case, perfect alignment of liability with responsibility, does not and cannot exist, because "government" and "industry," those entities most responsible, do not exist. Therefore, a cost-spreading strategy must be based on a second-best case. The real question is whether the inverse-proportion model of responsibility and liability actually depicts the second-best case or whether a better configuration exists.

In designing a mechanism for spreading abandonment costs, attention must be paid to two operations. First, the cost-spreading strategy must allocate past costs, resolving the abandonment problem. Second, the mechanism should be designed with an eye toward avoiding duplication of the problem in the future. If financial liability is to be more closely aligned with policy-making responsibility, those persons targeted as cost bearers should have either a greater role in policy making or a set of legal rules available to them for cost avoidance. We have seen that during the transition, mismatches and accommodations occur, thus setting the stage for regulatory reform.

Two fundamental reforms suggest themselves. The traditional model of responsibility/liability must be redefined: Either costs should be imposed up the line, for example, greater personal exposure by second-tier actors, or third-tier parties should more directly participate in decision making. Either choice starts to realign the old model, and either choice promises more fairness and better efficiency. The new model's element of greater participation is fairer, because those affected help determine their fiscal fate. The new model also should encourage efficiency by discouraging bad investment decisions. If nuclear regulation is to respond satisfactorily to the abandonment-costs problem, and is to be receptive to a more sound nuclear policy, then the regulatory system must realign financial liability with policy-making responsibility in a way that promotes fairness and efficiency.

7. Transition and Beyond

LAW FOLLOWS POLICY

The promotion of nuclear power by government and industry has cost the United States billions of dollars and has produced little electricity for that investment. The political decision to go nuclear was aided and abetted by legal institutions and rules that now act as a continuing drag on responsive regulatory reform efforts. In an activist state, that is, in a state with an interventionist and bureaucratic governmental structure, law is intended to implement political preferences. Pluralist democratic theory, together with the doctrine of separation of powers, and the system of checks and balances allow accommodations among competing values and foster the expression of a multitude of voices throughout the polity. The fact that law helps facilitate policy choices is not without its downside. If the political climate or the market demand changes, then the policy preference will change, as well. However, since the policy preference has been embedded in society by law, the policy is stuck in the system and influences policy development. With the case of nuclear power, the policy was to promote nuclear-generated electricity. The policy was assisted by federal agencies promoting the licensing of nuclear facilities, Congresses willing to limit industrial liability, a ratemaking regime interested in capital expansion, and a consuming public desirous of using cheap electricity. The embeddedness of the promotional policy has produced the series of accommodations being made to salvage the industry.

The single problem of nuclear plant cancellations has been used as a staging area to question the configuration of industry-government relations in a high-technology world; the changing nature of the market for nuclear-generated electricity; the viability of traditional ratemaking; and the current dislocation between legal liability rules and fiscal responsibility. This wide-ranging discourse is held together by a core concept: Nuclear power and its regulation are being transformed from a traditional capital-expansionist model to a late capitalist postindustrial model. This chapter discusses the implications of the transition period for the development and interaction of nuclear law and policy. Indeed, the conceptual background of the book posits that law and policy have a peculiar interaction in the modern bureaucratic state. Recognizing that interaction is necessary for understanding policy formation and for regulatory reform.

The interaction of law and policy can be expressed as a formula: In the modern bureaucratic state, law follows policy. A pristine legal culture of neutral decisional rules devoid of policy content can be conceived;[1] however, such a world is not our world and must remain imaginary. Law cannot help but reflect the economic, political, and moral values shaping policy. Indeed, law and policy interact in several configurations.[2] Legal cultures and subsystems within a single culture differ primarily in the degrees to which public policy influences law.[3] Modern government, with its emphasis on mid-term and long-term planning, depends heavily on law to implement policy.[4] Such dependency presents particular problems when policy choices turn out to be wrongheaded or mistaken or carry with them unforeseen consequences, as it has with the joint government-industry enterprise actively promoting the private commercialization of nuclear power. The unfolding story of nuclear power demonstrates that when there is a drastic change is external politics and markets, policy is splintered, and unfairness and inefficiency result.[5] Correcting these dislocations requires correcting established legal and policy-making institutions.

There are certain narrowing assumptions implicit in the "law-follows-policy" formula that must be stated. First, because defining *law* is the consuming task of jurisprudence,[6] and defining *policy* is that of political science,[7] working definitions of both are used. Second, the discussion of the interaction of nuclear law and policy is contextually based and historically bounded. The book's foreground focuses on domestic nuclear policy during an identifiable period starting a few seconds after 4:00 A.M. on March 28, 1979 and continuing to the present. The accident at Three Mile Island marks the end of the first stage of nuclear regulation and the beginning of the transitional period. Finally, the discussion of nuclear regulation also takes place within a distinct legal culture known as the activist state.[8]

These assumptions underlying the analysis of contemporary nuclear power regulation are primarily descriptive. There is also an important normative assumption operating. A pragmatic, but questioning, acceptance of the necessity of governmernt regulation is adopted. Sophisticated arguments can be made supporting the claim either that nuclear power is unnecessary for the country's energy program[9] or that bureaucratic regulation limits human action[10] and, therefore, nuclear regulation is unnecessary or wrong. These arguments are too utopian or too nihilistic to be taken seriously. They are unrealistic because they fail to recognize that government and industry have invested billions of dollars in nuclear technology over a period of several decades. Further, and more important, antinuclear and antibureaucratic arguments ignore or subvert the place of law in modern society. Law can contribute to a better and more just world even when wrongly influenced by policy, as is the case with nuclear regulation. The legal system is resilient enough to avoid forever staying mired in bad policy choices. Notwithstanding this affirmation of faith, regulatory breakdown occurs when governmental policy demands overload law, thus necessitating reform.

Having identified the assumptions, working definitions of the law-follows-policy formula must be provided. The terms *law* and *legal system* are used interchangeably. The legal system consists of decision-making structures, such as the Nuclear Regulatory Commission, and decision-making methodologies, such as ratemaking. When a structure applies a methodology, either to decide a past dispute or to issue a prospective promulgation, a rule results. *Law,* then, encompasses structures, methods, and rules.[11] *Policy* is more amorphously defined as the articulated political preferences and actions of government.[12] Further, policy is molded by two major forces—politics and markets.[13] The *activist state* is the historical continuation of and ideological successor to the New Deal. This conception of the state is grounded in the liberal ideal that in a complex world, central government must take an active role in mediating conflicts between the one and the many. Market imperfections mandate government involvement in the distribution and allocation of wealth and power based on principles of equity and fairness.[14] Equity and efficiency combine politics and economics in bureaucratic decision making, thus continuing Adam Smith's idea of political economy.

One refinement remains. Law and policy have a peculiar *interaction* in the context of the activist state. I have already said that in a modern capitalist democratic state, the legal system facilitates the policy choices of government. The promise of the law-follows-policy formula is that in the activist state, the legal system fairly and efficiently implements policy preferences.[15] Once public decision makers chose to promote nuclear power, the legal system was used to legitimize the choice. The downside consequence of law's playing handmaiden to policy is that policy preferences become institutionalized in the legal system. The risk is that once the policy preference sours, because political or market climates change, for example, policy choices linger in the legal system relied on, and in the case of nuclear power created, to promote a chosen policy. Thus, law and policy in an activist state interact, replicating policy choices in the recesses of the legal system. Nuclear regulation is experiencing the confusion of having an outworn policy trapped in a legal system of its own making. Now, as policy choices change, the legal system responds slowly (inefficiently) and oddly (unfairly), which results in regulatory failure.[16]

Nuclear policy before 1979 can be clearly identified. Government and industry combined in an effort to promote nuclear power as an increasingly important source of energy for the future. Early economic and political indicators were consistent with and supportive of this promotional policy. Nuclear power was safe, clean, cheap, and virtually inexhaustible. Public utilities, with a little persuasion, became convinced that they could realize economies of scale and could construct larger plants netting lower costs per kilowatt hour per construction dollar. Construction contractors and equipment vendors were developing services and products and expanding into new, and in some instances oligopolistic, markets. Consumers were to receive electricity "too cheap to meter." Further, the general public was behind the joint government-industry enterprise moving nuclear technology from military control into private commercialization. All of these factors combined to form a unified promotional policy.

Nuclear policy has become severely fragmented. The unified front no longer exists. Since the late 1960s, the environmental consequences of nuclear power have been consistently questioned. The economic advantage previously enjoyed by the industry has been rapidly reversed since the mid-1970s. Finally, the accident at Three Mile Island has radically altered public acceptance of the industry's safety claims. The public is no longer complacent about either the safety of nuclear plants or the costs associated with construction and operation. Divergence of opinion between government and industry concerning who should bear the cost of nuclear power indicates a significant readjustment in decision-making power. Additionally, industry squabbles concerning cost allocation are spreading with a proliferation of multi-million-dollar lawsuits among members of the atomic partnership. Shareholders are suing utilities, who are suing contractors, who are suing architect-engineers, who are suing equipment vendors.

The one stable nuclear policy variable is the vision of a nearly inexhaustible resource supply. This surfeit of energy potential, together with a multi-billion-dollar investment in plant, equipment, and services, means that nuclear power will not disappear from our overall energy planning. However, the shift from a monolithic promotional policy to an unstable fragmented picture has repercussions throughout society. There has been no investment in new nuclear plants since 1978, and none is forseeable absent a cataclysmic change in energy demand, economic markets, or world security. Further, over one hundred nuclear plants, many under construction, have been canceled at the cost of billions of dollars, which signifies a lost hope in nuclear energy.

Still, the early promotional policy, entrenched in the legal system, continues to influence nuclear regulation. During the formative period, government and industry insulated themselves from fiscal liability while willfully exercising virtually exclusive policy-making authority. Now, the market has caught up with the false economic assumptions blindly held by promoters, resulting in a fragmented nuclear policy and a breakdown between policy-making responsiblity and financial liability. Tens to hundreds of billions of dollars of costs associated with the mistaken decision to go nuclear must be paid and are earmarked to be absorbed by those persons least responsible for choosing the nuclear option. The impulse of the nuclear regulatory system is to have third-tier actors pay. These cost bearers are expected to pay for mistakes and receive no benefits.

Nuclear policy was embedded in law, and now law is called on to reform what has turned out to be an unfortunate policy. The thesis is that the transitional phase contains within it the seeds necessary for corrective action and reform. Policy makers can develop a more responsible nuclear policy by examining the causes of the breakdown of the former policy and by anticipating the consequences flowing from the existing set of legal rules and institutions established to further the promotional policy. The transition, thus, presents an opportunity for critique and reconstruction.

MODELS OF NUCLEAR REGULATION

By definition, transition means that the regulatory regime moves from one point to another. Nuclear regulation is moving and ought to move from a traditional to a postindustrial model of nuclear policy. I intend no ambiguity in the use of the descriptive *is* and the prescriptive *ought to*. These are not discrete phenomena. What nuclear policy *is* during the transition is also a part of what that policy will and *ought to* become when it is reformulated. Naturally, the new policy will take some of the characteristics of the old. Figure 7-1 outlines the characteristics of the regulatory models used throughout this book.

The models should now be familiar. The traditional model, promoting nuclear power, defines the hard energy path. The model conceives of energy supply as large-scale, high-technology, capital-expansive, and necessary for industrial and economic growth and social well-being. The regulatory structure attached to this model is, characteristically, elitist, specialized, centralized, and technocratic, with little need for public participation. Public participation is viewed as inefficient. Instead, expert federal administrative agencies are the primary decision makers. The economic-incentive structure in this model comes in the form of direct government support through legislation that gives private industry access to nuclear materials, invests in research and development, and limits the financial exposure of participants in the nuclear venture. The consequence of this model is the creation of a nuclear market that would not otherwise exist. The nuclear market is an artificial construct and is susceptible to destabilization when political and

Figure 7-1. Models of Nuclear Policy and Regulation

	TRADITIONAL	TRANSITIONAL	POSTINDUSTRIAL
Dates	1946-1979	1979-present	
Energy supply	high-technology, large-scale	mixed	competitive, alternative sources
Plant construction	capital-expansive	stalled, canceled	scaled-down, standardized
Decision-making authority	centralized, federal	dualistic, federal/state	decentralized federal/state/local
Bureaucratic model	elitist, expert	adversarial, contentious	democratic
Decision-making fora	agencies (NRC, PUCs, DOE)	courts	community-based bargaining (e.g., negotiation, mediation)
Ratemaking	rate base cost-based	accommodationist	market-based, deregulation
Public participation	circumscribed	contentious, factious	participatory
Policy support		diffuse	pluralistic

economic structures change. The destabilization has occurred as the financial market ceases investing and as public support decreases, thus eroding the legitimacy of the traditional model.

The transitional model is less a model than a state of temporary destabilization. Energy supply during this period continues, naturally enough; however, the traditional sources of supply are questioned which results in the delay and cancellation of nuclear plants. The commitment to large-scale, high-technology plants is on hold, particularly as the energy supply is supplemented by cogeneration, self-generation, and alternative sources. Decision making is decentralizing, moving from the federal to state governments. Further, agency actions are increasingly second-guessed and overturned or delayed by courts. This shift to courts evinces a breakdown in political support as not only intervenors but also shareholders, utilities, contractors, vendors, and architects and engineers seek judicial relief from the financial consequences of nuclear market failure. The shift also demonstrates dissatisfaction with elitist bureaucratic decision making. If the public voice is not heard at the agency, then the court is another forum.

During the transitional period, the pronuclear incentive structure is also questioned. Government subsidy is viewed as dangerous, and the traditional ratemaking formula proved inadequate to handle the allocation of plant cancellation costs. The result is a series of regulatory accommodations imposing losses among consumers, shareholders, and taxpayers. The accommodationist response is built on a shaky economic foundation and on conflicting policy choices, resulting in cost-allocation decisions that do not uniformly honor the connection between policy-making responsibility and financial liability.

If the decentralization shift in decision-making authority from agencies to courts and from the federal government to state legislatures is evidence of political change, then there is a concomitant economic change as the rising price of electricity from large-scale plants meets with increased competition. The artifice of the nuclear market is exposed by the substitution of electricity from other suppliers, thus replacing the state-created nuclear market with a less regulated, more competitive market.

The postindustrial model relies on electricity supplied from several sources. Central power stations no longer serve as the sole supplier. For nuclear power, that means that scaled-down, standardized plants rather than mega-plants stand a better chance of competing against other suppliers. The bureaucratic mode of decision making will be democratic and participatory, rather than technocratic and elitist. This reformulation of the role of agencies does not mean that experts have no place in administrative decision making. Rather, experts are used to gather and to interpret positive scientific data, and politically accountable decision makers make policy recognizing that normative uncertainty pervades nuclear decision making. Correlative with the change in bureaucratic thinking is a move toward a wider range of dispute-resolution alternatives. In addition to courts and agencies, community-based and interest-based bargaining will take place utilizing negotiation, mediation, and arbitration as viable alternatives to litigation and administrative hearings.

Finally, the nuclear market in the postindustrial model is more competitive and is less supported by the state than either of the other models. While the state will continue to monitor safety, financial matters will not be subsidized by the state. Price-Anderson Act limitations, for example, will be raised or eliminated, and rates will be based on the economic value of a kilowatt, not the utility's production costs.

These three models describe the characteristics of regulatory periods. We are currently experiencing a transition, and a closer examination of the transitional model reveals the contours of future nuclear law and policy.

DUALITIES AND THE TRANSITIONAL MODEL

Transitions are inherently discontinuous as a movement is made from one period (or model) to another. Throughout the book, a series of dualities were mentioned. The dualities represent the discontinuities present during transition. Thus, the transition is exemplified by the presence of dualities. It also appears that the dualities are, for the most part, mutually exclusive. Further, the dualities themselves are discontinuous. That is to say, the dualities, without being forced, do not align themselves perfectly with either the traditional model or the postindustrial model. Instead, they exist for, or more accurately were created by, decision makers as justifications for their decision making. Thus, the dualities, listed in figure 7-2, serve the heuristic purpose of exposing the rhetorical devices and argumentative strategies used during the transition. The existence of discontinuous dualities is also significant because decision making during transition is marked by its contentious and adversarial nature. In short, decision makers pick and choose from mutually exclusive, competing policy arguments rather than base decisions on a clearly articulated policy.

Although figure 7-2 aligns the dualities, the alignment is forced. Decision makers pick and choose from among the dualities as justifications for their decisions. The traditional model, for example, favors long-term efficiency and competitive pricing or market mimicking as ways to maximize wealth and grow economically. Instead, during the transition, the traditional ideology, which supports the continuance and support of large central power stations, must achieve its ends through short-term loss allocation based on ex post argumentation. Similarly, proponents of the postindustrial ideology, who favor short-term equity, avail themselves of long-term efficiency arguments during the transition. Both postures mix ideological elements and are consistent with a radically changing market structure, a lack of well-developed theory, and inconsistent decision making.

A pure version of the traditional and postindustrial models would align the dualities as they are listed above. The traditional model would treat nuclear policy from an ex ante perspective and would formulate a long-term, systematic, nuclear program in order to maximize efficiency and economic growth. The traditional model would also consist of elite bureaucratic decision making with minimal public participation and would justify shareholder protection as a means to wealth-maximizing ends. Likewise, the postindustrial model would attempt to create a policy motivated by political participation and would concern itself with short-term, project-specific decentralized decision making. Equity would serve as the decisional rule for the ex post protection of consumers, those persons in the worst position to protect themselves.

LESSONS OF THE TRANSITION

The transitional dualities betray a lack of theory for cost allocation, indicate a change in direction for nuclear policy, and expose the values missing from the traditional model to reveal the configuration of the postindustrial model. Because the transitional period discloses the weakening of the traditional

Figure 7-2. Transitional Dualities

	Traditional Model	Postindustrial Model
Goals and objectives	long-term efficiency economic growth	short-term equity political participation
Bureaucratic model	managerial	democratic
Argumentative perspectives	ex ante quantitative	ex post nonquantitative
Constituency	shareholders	consumers
Public participation	mechanistic	pervasive
Cost-allocation perspective	systemic	project-specific

model and prefigures the postindustrial model, we can see the dynamic quality of law and policy. Further, the dualities serve to critique the traditional model and present an opportunity to reconstruct a more responsive regulatory approach. This backwards-forwards examination of nuclear law and policy should prove sturdy enough to suggest regulatory reform, because future reforms emanate from past practices, thus providing stability between regulatory regimes.

Developing a list of concrete reform proposals is both easier and harder to write about than might be suspected. It is easy to invent specific reforms, because they are limitless. It is hard to deal with specific reforms, because what must be broken is a forty-year-old mindset favoring nuclear power, and a forty-year investment in a now-failing enterprise. Without more, concrete reforms will not break that mindset.

Reform and reconstruction, then, must be grounded in the lessons of the transitional period. The lessons are derived from the critique of the former model and the assessment of the current situation. The information derived from the critique and assessment is necessary but not sufficient for the purpose of formulating concrete reform proposals. In addition to critique and assessment, the study of the transition requires an explication of the values contained in a more responsive regulatory regime. Until these issues are addressed at a normative level, no reform effort will work.

The traditional model broke down because it curtailed participation and attenuated responsibility in decision making. Further, nuclear policy during this period depended, unwisely, on an artificial market of its own creation and highly centralized decision making, thus aggravating the disparity between responsibility and liability. Simply, the development of nuclear power stalled because of the regulatory system cut itself off from the market and was not sensitive to politics. Recognizing the underlying values of participation, decentralization, competition, and responsibility is imperative for sound regulatory reform.

The method of ascertaining the regulatory future simply aligns the mutually exclusive dualities, describes their consistencies, generates sets of reforms, decides on a way of choosing between sets, chooses between sets, and then designs concrete reforms consistent with the chosen set. The evaluative test for choosing between opportunity sets is responsiveness: Does the chosen reform set respond to the critique? If the choice is wisely made, then legitimacy follows.

The opportunity sets are clear. The choice is between the traditional model of elitist, technocratic, nuclear regulation and the postindustrial model of participatory, decentralized regulation. These models are mutally exclusive ways of looking at the world. They are regulatory ideologies. The traditional model is a vision of an end state where nuclear power necessarily plays a part in our energy future. The postindustrial model is a state of second-best. Nuclear power may or may not be part of the postindustrial world as more than one energy policy competes for our attention. Nuclear power may play a role if political and market circumstances are such that politically responsive and accountable decision makers choose nuclear through democratic participatory processes.

Policy options under the postindustrial model reflect tensions between politics and markets, power and money, and equity and efficiency. The competing policies are given voice through decision-making processes that require participation before legitimacy attaches. The outcome and future depend less on a vision of the best end state than on processual decision making and policy making that yield a realignment of responsibility and liability through responsive regulation. If nuclear power reemerges, the reemergence will be the result of public choice, not industry-government edict.

This book is a critique of the traditional model, an assessment of the discontinuities existing during the transition, and an explication of the norms that should be present in the new regulatory structure. So conceived, the book is a necessary prelude to a more elaborate discussion of nuclear regulatory reforms.

Concrete reforms emanate from the critique and assessment. Specific examples of reform at the federal level would include: abolishing the NRC; creating an independent safety body; establishing a separate licensing agency with a separate appeals board or eliminating administrative appeals; relying on "hard look" judicial review; repealing the Price-Anderson Act; and creating a federal Office of Public Advocate, which would provide expertise and funding for intervenors. At the state level, concrete reforms would include: restructuring the ratemaking formula to avoid cost-plus pricing; extending legislative ventures into the nonradiological side of nuclear regulation; funding intervenors; and encouraging interstate and state-federal participation in such issues as waste disposal, nuclear transportation, and emergency planning. These proposals, standing alone, are not novel and are easily generated.[17] These reforms are consistent with the critique because they are based on increased participation and increased decentralization aimed at a closer alignment of responsibility and liability. As such, they coincide with the postindustrial model.

An alternative set of reforms, consistent with the traditional model, can also be fashioned. At the federal level, traditional model reforms would include: streamlining licensing as per the NRC task force report; renewing Price-Anderson; circumscribing public participation; supporting standardization financially; and promoting the development of a national power grid. At the state level, reforms would encompass: continued use of cost-based ratemaking; giving greater judicial deference to PUCs; lowering the standard for prudency findings; eliminating proxy advocates; and increasing the expertise of PUCs by appointing commissioners and expanding technical staff.

These sets of proposals are held together by normative values. One set conforms with the postindustrial model, and the other with the traditional model. The choice between these two opportunity sets should be clear. Given the changes in the industry, choosing the traditional model is the more radical, and less intelligent, choice. The traditional reforms are not responsive to political or market changes and, thus, cannot promise either greater efficiency or fairness. Postindustrial reforms, then, are more responsive to political and market developments.

CONCLUSION

There are clear political implications in treating the transitional period as a bridge between two models and as an opportunity for policy development. So viewed, the transition is the dynamic part of bureaucratic government during which corrective action takes place, thus relegitimizing policy choices. Even when the transition is rough, it serves to undo the fetters of past, and here mistaken, policy that has been institutionalized in society by law.

The implication of the transition for the interaction of law and policy is that allegiance to unitary or end-state theories of law or policy is unnecessary. Instead, a policy-making process, and the role and rule of law in that process, are recognized as constitutive of a polity that attempts to mediate conflict and give voice to varied interests rather than dominance to some. Processural mediation of competing claims, then, becomes the hallmark of legitimate policy making and decision making in the modern state. Democratic and pluralistic participation is accorded priority as both economic and political values coexist and compete for policy making consistent with the American liberal traditional of law and politics.[18]

The inescapable consequence of the interaction of law and policy is that political values are implicated in the regulatory order. It can be no other way. The legitimacy of the regulatory state depends on acceptability and accountability. As long as the public accepts the decisions of the state and decison makers maintain their accountability to their various constituencies, stability results. Destabilizing events can occur. The rejection of a promotional nuclear power policy by financial markets in the mid-1970s and the disenchantment of the general public after TMI are prime examples of such events. Destabilization, however, need not be permanent, and the country need not experience a continuous legitimation crisis as long as transition during which reforms responsive to the assessment of the destabilizing events and to the critique of past policy can occur.

The primary significance of the transition is that the radical changes in politics and markets produced a series of mismatches in nuclear policymaking. The promotional policy met with serious antagonism. During the transition, accommodations were made through bargaining and compromise. The accommodations are not held together by a coherent economic model or a clear policy preference. Instead, they represent a temporary "balancing" of interests. The transitional balancing allows the legal/policymaking system to shake out wrinkles (discontinuities) and allows policy paradigms to transform and reconstruct. The existence, purpose, and effect of the transitional model are consistent with government in a pluralistic democracy and are consistent with the traditional liberal theory of law and politics. Transition, then, is a necessary part of the activist bureaucratic state.

NOTES

Introduction

1. The description of the nuclear accident at Three Mile Island is taken from Daniel Ford, Three Mile Island: Thirty Minutes to Meltdown 16–34 (1982), and Report of the President's Commission on the Accident at Three Mile Island (Washington: October 1979), also known as the Kemeny Commission Report.

2. A "common-mode" failure as described by Daniel Ford, id. at 13:

> A cardinal rule for the designers of commercial nuclear power plants is that all systems essential to safety must be installed in duplicate, at least, so that if some of the apparatus fails, there will always be enough extra equipment to keep the plant under control. Federal regulations governing the industry require strict conformity to this prudent design philosophy. Even when this rule is applied, however, there is a type of accident that can jeopardize the safety of a nuclear plant. This type of accident involves what is known as a common-mode failure—a single event or condition that can cause simultaneous multiple malfunctions resulting in the major disruption of the plant's safety system.

The idea that an unplanned occurrence can overcome multiple precautions typifies not only nuclear plant technology but nuclear regulation, as well. TMI unwittingly began a new era in commercial nuclear policy sensitive to the fact that no fail-safe system exists.

3. William Lanouette, in Atomic Energy, 1945–1985, Wilson Quarterly, vol. 9, p. 9 (Winter 1985), argues that TMI was merely the culmination of a nuclear power program fated for trouble from its inception. I concur in that assessment. Still, TMI exists as a clear and remarkable signal of the end and beginning of commercial nuclear eras.

4. These figures are based on what is popularly referred to as the Rasmussen Report, Nuclear Regulatory Commission, Reactor Safety Study: An Assessment of Accident Risk in U.S. Commercial Nuclear Power Plants, ch. 5 Wash-1400, NUREG 75/014 (Washington: Nuclear Regulatory Commission October 1975). The Rasmussen Report was heavily criticized, and another study was undertaken. Although the methodologies employed by the Rasmussen Report were accepted, the statistics were partially disapproved by the NRC in the Lewis Committee Report, Nuclear Regulatory Commission, Risk Assessment Review Group Report, NUREG/CR-0400 (Washington: Nuclear Regulatory Commission 1978). See also Nuclear Energy Policy Study Group, Nuclear Power Issues and Choices, ch. 7 (Cambridge: Ballinger Publishing Co. 1977).

5. Ford, note 1 at 33.

6. Diamond, Problems Delay Three Mile Island Work, N. Y. Times, July 25, 1984, p. 12, col. 1. Douglas Bedell, TMI communications manager for General Public Utilities, the owners of TMI, confirmed the $1 billion figure in a telephone interview. He also noted that this amount was limited to clean-up costs and did not include such items as cost of replacement power.

7. Plant decommissioning can be accomplished in one of two ways. The obsolete plant can either be buried in concrete or dismantled. NRC regulations for decommissioning are contained in 10 C.F.R. §§50.33(g) and 50.82 (1984). Early estimates were that plant decommissioning would cost $50–100 million. Because of the lack of a present solution to the waste disposal problem, these costs rival or exceed construction costs. John Ferguson, Preston Collins, Nicholas Reynolds, Financial Aspects of Power Reactor Decommissioning in Iowa State University Proceedings by Regulatory Conference 471 (1980); and Buta and Palmer, An Analysis of Decommissioning Cost Estimates for Nuclear Operating Plants, 114 Public Utilities Fortnightly 47 (July 19, 1984).

8. The TMI accident has generated litigation in federal and state courts and in federal and state agencies. A sampling includes: Commonwealth of Pennsylvania v. General Public Utilities Corp., 710 F.2d 117 (3d Cir. 1983) (although private litigants cannot sue under the Atomic Energy Act, state law is available); General Public Utilities Corp. v. Babcock & Wilcox, 547 F. Supp. 842 (S.D.N.Y. 1982) (TMI owners sue reactor manufactors, suit settled); Susquehanna Valley Alliance v. Three Mile Island Nuclear Reactor, 485 F. Supp. 81 (M.D. Pa. 1979) (plaintiffs in environmental suit had to bring claims first to NRC rather than federal court); Metropolitan Edison Co. v. People against Nuclear Energy, 460 U.S. 766 (1983) (NRC is not required to assess psychological damages as part of its environmental assessment); and Pennsylvania Electric Co. v. Pennsylvania Public Utility Commission, 467 A.2d 1367 (Pa. Cmwlth. 1983) (PUC ruling that shutdown of nuclear power station not used and useful).

The Price-Anderson Act, to be discussed in chapter one, establishes an insurance pool for victims of nuclear incidents. Approximately $28 million of the $30 million spent from the fund since 1957 was allocated to TMI claims in the following allotments: $1.4 million for living expenses to families with pregnant women and preschool children who evacuated the area at the governor's suggestion; $20 million for economic harm to businesses and individuals; $5 million for a public health fund; and $2.5 million for attorneys' fees and expenses. Nuclear Regulatory Commission, The Price-Anderson Act—The Third Decade, NUREG-0957 (Washington: Nuclear Regulatory Commission December 1983) at I-6 to I-7.

9. Munson, The Price Is Too High, in Robert Engler (ed.), America's Energy 39 (New York: Pantheon 1980).

10. These figures were compiled by the Union of Concerned Scientists in a pamphlet titled Nuclear Power Plants in the United States: Current Status and Statistical History (Cambridge: Union of Concerned Scientists May 1, 1984). The information was gathered from the Nuclear Regulatory Commission, the Department of Energy, and the Atomic Industrial Forum. Office of Technology Assessment, Nuclear Power in an Age of Uncertainty, ch. 3, OTA-E-216 (Washington: February 1984) (hereinafter referred to as OTA Report); Department of Energy, U.S. Commercial Nuclear Power: Historical Perspective, Current Status, and Outlook 16–17, DOE/EIA-0315 (Washington: Department of Energy March 1982), and Department of Energy, Nuclear Plant Cancellations: Causes, Costs, and Consequences 3–17, DOE/EIA-0392 (Washington: Department of Energy April 1983).

11. OTA Report, note 10 at 3. The assertion that the nuclear problem is temporary is an attempt at realism that does not examine a nonnuclear scenario. The future of nuclear power must be assessed in two ways. The first generation of power plants, those from the early 1960s to the present, will continue to need maintenance and supervision, and several dozen plants remain to be brought on line. There is no second generation of plants in the forseeable future. We cannot predict when new plants will be ordered absent a major alteration in energy planning such as unanticipated growth in electrical demand, a major energy supply disruption, a mandate

against coal-fueled plants, nuclear plant standardization with preapproved sites, or fuel-regulating reform. A nonnuclear future is a possibility that is mentioned only in passing. For example, see Union of Concerned Scientists, A Second Chance: New Hampshire's Electricity Future as a Model for the Nation (Cambridge: Union of Concerned Scientists 1983).

12. Peter Shuck, When the Exception Becomes the Rule: Regulatory Equity and the Formulation of Energy Policy through an Exceptions Process, 1984 Duke Law Journal 163, and Joseph P. Tomain, Institutionalized Conflicts between Law and Policy, 22 Houston Law Review 661 (1985).

13. Fortune Magazine, The Five Hundred: The Fortune Directory of the Largest U.S. Industrial Corporations at 276–78 (April 30, 1984).

1. Institutional Setting

1. Histories of the regulation of the nuclear industry are contained in: John Chubb, Interest Groups and the Bureaucracy: The Politics of Energy, ch. 4 (Stanford: Stanford University Press 1983); David Davis, Energy Politics, ch. 6 (New York: St. Martin's Press 1974); Alfred Aman, Energy and Natural Resources Law: The Regulatory Dialogue §7.01-.03 (Charlottesville: Michie Bobbs-Merrill 1983); Donald Zillman and Lawrence Lattman, Energy Law 582–96 (Mineola: Foundation Press 1983); James Quirk and Katsuaki Terasawa, Nuclear Regulation: An Historical Perspective, 21 Natural Resources Journal 833 (1981); Irvin Bupp and Jean-Claude Derain, Light Water: How the Nuclear Dream Dissolves (New York: Harper and Row 1978); and George T. Mazuzan and J. Samuel Walker, Controlling the Atom: The Beginnings of Nuclear Regulation 1946–1962 (Berkeley: University of California Press 1984).

2. Freeman Dyson, Disturbing the Universe 51–53 (New York: Harper Colophon 1979).

3. Diane Maleson, The Historical Roots of the Legal System's Response to Nuclear Power, 55 Southern California Law Review 597, 600 (1982).

4. Pub. L. No. 79-585, 60 Stat. 755 (1946).

5. Pub. L. No. 83-703, 68 Stat. 919 (1954).

6. Nelson Polsby, Political Innovation in America 18-35 (New Haven: Yale University Press 1984).

7. H.R. Rep. No. 2181, 83d Cong., 2d Sess 1 and 3 (1954).

8. Department of Energy, U.S. Commercial Nuclear Power: Historical Perspective, Current Status, and Outlook, DOE/EIA-0315 (Washington: Department of Energy March 1982) at 6.

9. Pub. L. No. 85-256, 71 Stat. 576 (codified as amended at 42 U.S.C. §82210 [1982]).

10. Mark Hertsgaard, Nuclear, Inc. 33 (New York: Pantheon, 1983).

11. Nuclear Regulatory Commission, The Price-Anderson Act—The Third Decade, NUREG-0957 (Washington: Nuclear Regulatory Commission December 1983).

12. Guido Calabresi and Phillip Bobbitt, Tragic Choices (New Haven: Yale University Press 1978).

13. Duke Power Co. v. Carolina Environmental Study Group, Inc., 438 U.S. 59 (1978) (Supreme Court holds Price-Anderson Act constitutional). The act has been amended twice, once in 1967, then again in 1977 with slight modifications. As the act comes up for renewal in 1987, debate about the how large a cushion the industry should have has started. H.R. 421, 98th Cong., 1st Sess. (Jan. 3, 1983) titled the Nuclear Incident Liability Reform Act of 1981, which is intended to increase industry's liability for accidents.

14. Christopher Flavin, Nuclear Power: The Market Test 3 (Washington: Worldwatch 1983).

15. Hertsgaard, note 10 at 44.

16. Office of Technology Assessment, Nuclear Power in an Age of Uncertainty, OTA-E-216 (Washington: Department of Energy February 1984) at 29–30; and Richard Pierce, The Regulatory Treatment of Mistakes in Retrospect: Canceled Plants and Excess Capacity, 132 University of Pennsylvania Law Review 497, 500-02 (1984).

17. The claim that nuclear was cheaper than coal was demonstrably true prior to TMI. Compared with a coal-fired unit, nuclear was a cheaper plant to build and to operate. This assertion is no longer the case. Office of Technology Assessment Report, note 16 at 57-71, and Charles Komanoff, Assessing the High Costs of New U.S. Nuclear Power Plants (June 1984) (paper on file with author).

18. U.S. Commercial Nuclear Power, note 8 at 9.

19. Bupp and Derain, note 1 at 49.

20. U.S. Commercial Nuclear Power, note 8 at 10 table 1.

21. 42 U.S.C. §§4321-70 (1982).

22. Calvert Cliff's Coordinating Committee, Inc. v. United States Atomic Energy Commission, 449 F.2d 1109 (D.C. Cir. 1971).

23. 10 C.F.R. Part 51 (1984).

24. The 1964 amendments included Pub. L. No. 88-489, 78 Stat. 602 (1964), titled the Private Ownership of Special Nuclear Materials Act, which gave private industry more control of nuclear fuel. Once a plant is built, it must be serviced and supplied with fuel over the thirty- to forty-year, or more, life of a plant. Fuel service in itself is a multi-billion-dollar industry.

25. Energy Reorganization Act of 1974, P.L. 93-438, 88 Stat. 1233 (1974) (codified as amended at 42 U.S.C. §§5801–79 [1982]).

26. Power Reactor Development Co. v. International Union of Elec., Radio & Mach. Workers, 367 U.S. 396 (1961).

27. International Union of Elec., Radio & Mach. Workers v. United States, 280 F.2d 645 (D.C. Cir. 1960).

28. 367 U.S. at 408.

29. 367 U.S. at 415–16.

30. Peter Strauss, The Place of Agencies in Government: Separation of Powers and the Fourth Branch, 84 Columbia Law Review 573 (1984).

31. A constant debate exists among scholars and within the judiciary on the scope of judicial review of administrative action. Compare Harold Leventhal, Environmental Decisionmaking and the Role of the Courts, 122 University of Pennsylvania Law Review 509 (1974), with David Bazelon, Coping with Technology through the Legal Process, 62 Cornell Law Review 817 (1977), for the views of two distinguished jurists on the question of how much deference is due agencies in highly technical areas. William Rodgers, A Hard Look at *Vermont Yankee* Environmental Law under Close Scrutiny, 67 Georgetown Law Journal 699 (1979), and William Rodgers, The Natural Law of Administrative Law, 48 Missouri Law Review 101 (1982). Professor Joel Yellin argues that courts are not equipped to second-guess agencies in the area of nuclear power and that special masters might perform the task better, in High Technology and the Courts: Nuclear Power and the Need for Institutional Reform, 94 Harvard Law Review 489 (1981). Professor Yellin has also suggested the creation of an executive-legislative hybrid institution through which techno-scientific policy positions can be recommended to Congress, as well as a science advisory board for the judiciary, Joel Yellin, Science, Technology, and Administrative Government: Institutional Design for Environmental Decisionmaking, 92 Yale Law Journal 1300, 1326–28 (1983).

32. 5 U.S.C. §706 (1982).

33. 447 F.2d 1143 (8th Cir. 1971) *aff'd mem.* 405 U.S. 1035 (1972).

34. 447 F.2d at 1154.

35. Justices Douglas and Stewart dissented from the affirmance. Summary affirmances are "not to be read as an adoption of the reasoning supporting the judgment under review." Zobel v. Williams, 457 U.S. 55, 64 n.13, 102 S. Ct. 2309, 2315 n.13 (1982); and Mandel v. Bradley 432 U.S. 173, 176 (1977); Pacific Gas & Electric Co. v. State Energy Resources Conservation and Development Comm'n, 461 U.S. 190 (1983).

36. 461 U.S. 190 (1983).

37. Radiation levels and half-life expectancies depend on the nature of the nuclear by-product. A contaminated reactor, for example, can emit low-level radiation with negligible health or environmental effects. Plutonium, at the other extreme, is the most highly toxic element known and has a half-life of tens of thousands of years. Harold Green and Clifford Fridkis, Radiation and the Environment, in Federal Environmental Law 1022 (Erica Dolgin and Thomas Guilhert eds.) (St. Paul: West Publishing Co. 1974); Nuclear Energy Policy Study Group, Nuclear Power Issues and Choices, ch. 5 (Cambridge: Ballinger Publishing Co. 1977); Dean Hansell, The Regulation of Low-Level Nuclear Waste, 15 Tulsa Law Journal 249 (1979); Lash, A Comment on Nuclear Waste Disposal, 4 Journal Contemporary Law 267 (1978); Richard Ausness, High-Level Radioactive Waste Management: The Nuclear Dilemma, 1979 Wisconsin Law Review 707.

38. See City of West Chicago v. NRC, 701 F.2d 632 (7th Cir. 1983); Potomac Alliance v. NRC, 682 F.2d 1030 (D.C. Cir. 1982); and Lower Alloways Creek Twp. v. Public Service Electric & Gas Co., 687 F.2d 732 (3rd Cir. 1982).

39. Nuclear Waste Policy Act, 42 U.S.C. §§10101-226 (1982).

40. 42 U.S.C. §10132(b)(1)(B) (1982).

41. 42 U.S.C. §10132(b)(1)(C) (1982).

42. 659 F.2d at 923 quoting §274(k) of the Atomic Energy Act of 1954.

43. 659 F.2d at 925.

44. Maleson, note 3.

45. Note, The Supreme Court, 1982 Term, 97 Harvard Law Review 238, 242 (1983).

46. 42 U.S.C. §274(k).

47. Symposium, Federalism and Energy, 18 Arizona Law Review 283 (1976); Note, Nuclear Power Regulation: Defining the Scope of State Authority, 18 Arizona Law Review 987 (1976); Note, Pre-emption under the Atomic Energy Act of 1954: Permissible State Regulation of Nuclear Facilities, Location, Transportation of Radioactive Materials and Radioactive Waste Disposal, 11 Tulsa Law Journal 397 (1976); Note, Nuclear Waste Management: What the States Can Do, 1 Virginia Journal Natural Resources 103 (1980); Arthur Murphy and DiBruce La Pierre, Nuclear Moratorium Legislation in the States and the Supremacy Clause: A Case of Express Pre-emption, 76 Columbia Law Review 392 (1976); and Note, Application of the Pre-emption Doctrine to State Laws Affecting Nuclear Power Plants, 62 Virginia Law Review 738 (1976). Harvard professor Lawrence Tribe authored an influential law review article, California Declines the Nuclear Gamble: Is Such a State Choice Preempted? 7 Ecology Law Quarterly 679 (1979), which argued that California's moratorium legislation was based on economic, social, and psychological reasons that were not preempted. He successfully argued this position in Pacific Gas & Elec. Co. v. State Energy Resources Conservation and Development Comm'n, 461 U.S. 190 (1983).

48. Marshall v. Consumers' Power Co., 65 Mich. App. 237, 237 N.W.2d 266 (1975) (plaintiffs precluded from challenging plant's emergency core cooling system in state court); Commonwealth Edison Co. v. Pollution Control Board, 5 Ill. App. 3d 800, 284 N.E.2d 342 (1972) (state statute regulating radioactive discharge unconstitu-

tional); and New Jersey Dep't of Envir. Protection v. Jersey Central Power & Light Co., 69 N.J. 102, 351 A.2d 337 (1976) (state regulations in conflict with AEF regulations are preempted).

49. Stuart Diamond, The Heat Is Still Rising on Nuclear Regulators, N.Y. Times, July 29, 1984, p. 24E, col. 4; Matthew Wald, Panel Concedes Erring in Permit for Atom Plant, N.Y. Times, July 24, 1984, p. 7, col. 1; Union of Concerned Scientists and N.Y. Public Interest Group, The Indian Point Book (Cambridge: Union of Concerned Scientists 1982); and Testimony of Eric Van Loon and Ellen Weiss, The State of the Nuclear Industry and the NRC: A Critical View, before the Nuclear Regulatory Commission (Washington: Union of Concerned Scientists Nov. 17, 1983). The seductive safety-financial exchange is consistent with the currently popular bureaucratic methodology of cost-benefit analysis. An overreliance on cost-benefit data ignores fundamental, nonquantifiable assumptions such as: Is nuclear power worth the risk at any price? Risk assessment must be part of the overall decision-making process, and the costs of risks cannot be arbitrarily removed from more normative issues.

50. The connection between nuclear power and nuclear weapons is that the same technology is used to create nuclear fission. The enrichment of commonly mined uranium generates power and detonates bombs. As nuclear materials and technology have moved from government to private control, and as they are more liberally traded on world markets, the likelihood of weapons proliferation increases. Nuclear Energy Policy Study Group, note 37; and Amory Lovins and Hunter Lovins, Brittle Power: Energy Strategy for National Security, ch. 11 (Andover: Brick House Publishing 1982).

51. John Chubb, note 1 ch. 4.

52. Preemption cases in the nuclear field include: United States v. City of New York, 463 F. Supp. 604 (S.D.N.Y. 1978) (New York City precluded from licensing and regulating reactors in the city); Illinois v. General Electric Co., 683 F.2d 206 (7th Cir. 1982) cert. denied 461 U.S. 913 (1983) (Illinois Spent Fuel Act, which restricted disposal or storage of nuclear fuel held to be preempted by the Atomic Energy Act of 1954); United States v. State of Washington, 518 F. Supp. 928 (E.D. Wash. 1981) *aff'd* 684 F.2d 627 (9th Cir. 1982) cert. denied 103 S. Ct. 1891 (1983) (Washington's Radioactive Waste Storage and Transportation Act of 1980 preempted); United Nuclear Corp. v. Cannon, 553 F. Supp. 1220 (D.R.I., 1982) (Rhode Island's statute requiring the posting of a bond for waste disposal unconstitutional); and Silkwood v. Kerr-McGee, 464 U.S. 238 (1984).

53. An excellent description of the decade is Allen Matusow, The Unraveling of America: A History of Liberalism of the 1960's (New York: Harper and Row 1984).

54. Joseph Tomain, Institutionalized Conflicts between Law and Policy, 22 Houston Law Review 661 (1985); and Policy, Politics, and Law, 31 University of California Los Angeles Law Review 1247 (1984). Both articles discuss how energy policy is fragmented.

55. Historically, the states are the repositories of most ratemaking authority, and this subject will be discussed in detail in chapter 5.

56. Silkwood v. Kerr-McGee, 464 U.S. 238, 52 (1984). The Supreme Court's acknowledgment of the role of the states in the regulation of the nuclear power industry is also reflected in Van Dissel v. Jersey Central Power & Light Co., 465 U.S. 1001, 104 S. Ct. 989 (1984). In Van Dissel plaintiffs complained that the heat generated from a nuclear power plant warmed surrounding ocean water, causing worms to breed, which damaged their docks. Suit was brought in state court and dismissed because of the preemption doctrine. The U.S. Supreme Court denied certiorari, then granted certiorari, vacated the denial, and remanded for state court consideration in light of Silkwood. In the recent case of Cuomo v. Long Island

Office of the Special Counsel of the Merit Systems Protection Board 1 (Dec. 10, 1980).

10. Nuclear Regulatory Commission, Office of Inspection and Enforcement Region III: Report No. 50-358 80-09 (Washington: Nuclear Regulatory Commission July 3, 1980) at 6–13.

11. Note 9 at 11. Witnesses supported Applegate's charges with testimony about egregious safety violations; examples include: a radioactive waste drain clogged with concrete; several feedwater pumps filled with sand and mud; intake flues carrying makeup water to the cooling tower because of flaws in the design; 2,000-lb. fittings for residue head valves were installed where 5,000-lb. fittings were required; and a knowingly installed and ripped-out pipefitting at a cost of $320,000, at 13.

12. Government Accountability Project Request for Zimmer Construction Permit Cancellation, Letter Submitted to Region III, Nuclear Regulatory Commission, Mr. James Keppler (Director) (May 11, 1981) at 17.

13. Id. at 18.

14. Id. at 15.

15. Nuclear Regulatory Commission/Office of Inspection and Enforcement, Notice of Violation and Proposed Imposition of Civil Penalties, Dkt. No. 50-358, Construction Permit No. CPPR-88 EA-82-12 (Washington: Nuclear Regulatory Commission Nov. 24, 1981).

16. Id.

17. Nuclear Regulatory Commission, Report of the NRC Evaluation Team on the Quality of Construction at the Zimmer Nuclear Power Station, NUREG-0969 (Washington: Nuclear Regulatory Commission April 1983). The history of the NRC's safety evaluations of the Zimmer facility depicts how routine the evaluations were until the whistelblower Thomas Applegate successfully persuaded GAP to pursue the matter. The 1981 and 1982 safety evaluations are perfunctory and ignore the QA and managerial problems uncovered in 1983, even though those problems seemed to exist from the project's inception. Compare Nuclear Regulatory Commission, Office of Nuclear Reactor Regulation, Safety Evaluation Report, NUREG-0528 Supp. No. 2 (October 1981) and Supp. No. 3 (August 1982) with the 1983 evaluation.

18. Id. at 5.

19. Volume Two Torrey Pines Technology, Independent Review of Zimmer Project Management (GA-C17173, Aug. 1983) (hereinafter referred to as TPT Report).

20. Geraldine Brooks, Cincinnati Gas & Electric May Cancel Nuclear Plant in Face of Revised Costs for Finishing Unit, Wall Street Journal, Oct. 7, 1983, p. 6.

21. Zimmer, Preliminary Report of the Citizens' Advisory Committee, submitted to the Ohio Office of the Consumers' Counsel (April 18, 1984) §3.7.

22. Report of the President's Commission on the Accident at Three Mile Island 51–56 (Washington: October 1979). The NRC's preoccupation with promotion has deterred it from strenuous enforcement; for example, the Kemeny Report at 51 states: "We find that the NRC is so preoccupied with the licensing of plants that it has not given primary consideration to overall safety issues." See also Nuclear Regulatory Commission—The Rogovin Report, Hearings before a Subcommittee of the Committee on Government Operations, H.R. 96th Congr. 2d Sess, Feb. 13, 1983 at 9–50.

23. Complaint, note 7 at 22–25.

24. Order to Show Cause, note 8 at 4–6.

25. Id. at 12–13.

26. Id. at 14–17.

27. Nuclear Regulatory Commission, Report of the NRC Evaluation Team on the Quality of Construction at the Zimmer Nuclear Power Station, NUREG-0969 (Washington: Nuclear Regulatory Commission April 1983).

28. TPT Report, note 19 at 1–2.

Lighting Co., slip. op. (U.S. D.C. June 15, 1984) the court disallowed LILCO's preemption argument and allowed litigation to continue, which is indicative of the federal courts' new willingness to entertain state law claims in the nuclear area.

Since Silkwood, three court decisions indicate something of a reversion to deference to centralized control. Metropolitan Edison Co. v. People against Nuclear Energy, 460 U.S. 766 (1983); Baltimore Gas & Electric Co. v. Natural Resources Defense Council, Inc., 462 U.S. 87 (1983); and Florida Power & Light Co. v. Lorion, 105 S.Ct. 1598 (1985). For a longer exposition of federal court cases, see Joseph Tomain, Regulation in Transition in Progress in Nuclear Energy (1986) (forthcoming).

57. The Energy Emergency, 9 Weekly Compilation of Presidential Documents 1312–22 (Washington: Government Printing Office Nov. 7 and 8, 1973).

58. Tomain, note 54 and note 60 infra.

59. Department of Energy Organization Act, 42 U.S.C. §§7101–375 (1982); 1 Harold Green, Energy Law Service, ch. 2 (Chicago: Clark Boardman 1978); Alfred Aman, Institutionalizing the Energy Crisis: Some Structural and Procedural Lessons, 65 Cornell Law Review 491 (1980); and Clark Byse, The Department of Energy Organization Act: Structure and Procedure, 30 Administrative Law Review 193 (1978).

60. The Carter administration saw the promulgation of the National Energy Act of 1978 and the Energy Security Act of 1980, each of which contained major bills. The National Energy Act comprises five bills: the National Energy Conservation Policy Act, 42 U.S.C. §§8201–86b (1982); the Power Plant and Industrial Fuel Use Act, 42 U.S.C. §§8301–484 (1982); the Public Utility Regulatory Policies Act of 1978, Pub. L. 95-617, 92 Stat. 3117, Pub. L. 96-294, 94 Stat. 718, 770 (1978); the Natural Gas Policy Act, 15 U.S.C. §§3301–3432 and §7255 (1982); and the Energy Tax Act, Pub. L. 95-618, 92 Stat. 314, Pub. L. 96-223, 94 Stat. 273, Pub. L. 97-248, 96 Stat. 575 (1978). The Energy Security Act comprises the Defense Production Act Amendments of 1980, 50 U.S.C. §§2062–2166 (Supp. IV 1980); the United States Synthetic Fuels Corporation Act of 1980, 42 U.S.C. §§8701–95 (1982); the Biomass Energy and Alcohol Fuels Act of 1980, 42 U.S.C. §§8801–55 (1982); the Renewable Energy Resources Act of 1980, 16 U.S.C. §§2701–08 and 42 U.S.C. §§7371–75 (1982); the Solar Energy and Energy Conservation Act of 1980, 12 U.S.C. §§1451–1723 (1982); the Geothermal Energy Act of 1980, 30 U.S.C. §§1501–42 (1982); and the Acid Precipitation Act of 1980, 42 U.S.C. §§8901–12 (1982).

61. Department of Energy, Office of Policy, Planning, and Analysis, Energy Projections to the Year 2010, DOE/PE-0029/2 (Washington: Department of Energy October 1983) at 3-2, table 3-1.

62. Statistical Abstract of the United States at 578, table 986 (1984). Falling oil and natural gas prices in 1986 have the effect of making nuclear less attractive. At the writing of this book, the price decline is seen as temporary. The price story remains to be told.

63. Department of Energy, Nuclear Energy Cost Data Base 19, DOE/NE-0044 (Washington: Department of Energy October 1982). These figures are based on investment costs for the Northeastern United States. Other cost estimates also indicate that nuclear power has higher capital costs than does coal. Office of Technology Assessment, note 16 at 58–60; Lewis Perl, Estimated Costs of Coal and Nuclear Generation (Dec. 12, 1978) table 3; and Komanoff, note 17 at table 2. Charles Komanoff contrasts nuclear at $2,100/kwh with coal at $800–900/kwh.

64. Robert Stobaugh and Daniel Yergin, Energy Future (New York: Random House 1979).

65. Daniel Yergin and Martin Hildebrand, Global Insecurity (Boston: Houghton Mifflin 1982).

66. Robert Stobaugh and Daniel Yergin, note 64 at 10.

67. Many utilities are experiencing excess capacity rates of about 33%. Assuming that a reserve margin of 20% is reasonable, then the plant is 13% inefficient. How long the excess capacity will last is not easily determined. First, demand projections are difficult to assess. Second, demand elasticity in response to price increases is also difficult to ascertain. Third, the number of plants that will be retired and new ones that will come on line is not fixed. Office of Technology Assessment, note 16 at 31–45; Roger Colton, Excess Capacity: Who Gets the Charge from the Power Plant, 34 Hastings Law Journal 1133 (1983); and Pierce, note 16 at 511–17 and 538–41.

68. The DOE estimates abandonment costs at $8.1 billion in its worst-case study. Department of Energy, Energy Information Agency, Nuclear Plant Cancellations: Causes, Costs, and Consequences, DOE/EIA-0392 (April 1983) at xxi, table ES4. The DOE statistic is low even in its worst case. Economist Charles Komanoff estimates that the national economic damage caused by nuclear abandonment is $65–100 billion. This figure includes $15 billion invested in canceled plants; $20–40 billion in plants likely to be canceled; $30–40 billion attributable to plants with large cost overruns, Mark Hertsgaard, Nuclear Power: Too Costly to Save, N.Y. Times, June 24, 1984 p. F3, col. 1; and Komanoff, note 17.

69. See, e.g., Chemical Bank v. Wash. Pub. Power Sup. System, 666 P.2d 329 (Wash. 1983) (project participants not obligated to bondholders to continue with the project).

70. Marble Hill, Zimmer, and Shoreham are the subject of case studies in chapter two.

71. The General Accounting Office has issued a study that shows that the electric industry is enjoying renewed financial vigor except for utilities that are heavily invested in nuclear power. See General Accounting Office, Report to the Chairman, Subcommittee on Energy Conservation and Power, House Committee on Energy and Commerce, GAO/RCED-84-22 (June 11, 1984).

72. Hertsgaard, note 10.

73. Professor Richard Pierce has described the decision to construct a power plant as a "mistake in retrospect," by which he means that in some, if not many, instances the utility's decision to go nuclear was correct at the time it was made but would not be made today. Should a "mistake in retrospect" be treated differently than an initial mistake? This question will be discussed in chapter five. Pierce, note 16 at 498–99.

74. An interesting debate is developing between persons labeled separatists and nonseparatists who argue about whether it is possible to keep techno-scientific questions separate from legal-policy issues. Joel Yellin, note 31, 92 Yale Law Journal at 1305–18; and Stephen Carter, Separatism and Skepticism, 92 Yale Law Journal 1334 (1983). My sense is that this debate is a semantic quibble about labels and locating the gray area between the two sets of topics. Some things are scientifically accepted, e.g., the atomic weight of U-238. Others, such as the long-term effects of low-level radiation, are not scientifically known and are "policy" matters. I favor a system that encourages interaction between these sets rather than ignoring the gray area. Questions about risk are essentially mixed fact-value questions, Baruch Fischoff et als., Acceptable Risk (Cambridge: Cambridge University Press 1983). The separatist/nonseparatist dichotomy is also prevalent in administrative law scholarship, Gerald Frug, The Ideology of Bureaucracy in American Law, 97 Harvard Law Review 1276, 1298: "Values, ends, and desires—the subjective part of the human personality—are the attributes of the constituents who control the bureaucracy rather than the bureaucracy itself."

75. A "polycentric" decision is described by Lon Fuller:

> We may visualize this kind of situation by thinking of a spiderweb. A pull on one strand will distribute tensions after a complicated pattern throughout the web as a whole. Doubling the original pull will, in all likelihood, not simply double each of the resulting tensions but will rather create a different complicated pattern of tensions.

Lon Fuller, The Forms and Limits of Adjudication, 94 Harvard Law Review 353, 395 (1978). Milton Wessel, Science and Conscience 4–10 (New York: Columbia University Press 1980) (Professor Wessel calls these complex problems "socio-scientific" disputes); and Yellin, note 31 at 495–508. Nuclear issues not only are technically and scientifically complex, they also contain competing normative (political, philosophic, economic, and social) uncertainties. Robert Goodin, Political Theory and Public Policy, ch. 10 (Chicago: Chicago University Press 1982). In their work on risk decisions, the authors of Acceptable Risk, note 74 at 9, identify five "generic" complexities:

> (a) uncertainty about how to define the decision problem, (b) difficulty in assessing the facts of the matter, (c) difficulties in assessing the relevant values, (d) uncertainties about the human element in the decisionmaking process, and (e) difficulties in assessing the quality of the decisions that are produced.

76. See Fischoff et als., note 74 at xiii:

> Hence choosing an approach is a political act that carries a distinct message about who should rule and what should matter. The search for an objective method is doomed to failure and may obscure the value-laden assumptions that will inevitably be made.

2. *Three Failed Decisions*

1. The book assumes that regulation is necessary in this field. There never ha been a no-regulation or free-market approach to nuclear power, in part because c industry's desire to have federal support. Another reason is that the free marke cannot allocate nuclear power's externalities, which include health, safety, and en vironmental risks, and, now, abandonment costs.

2. Ben Kaufman, Zimmer: The Implications, Cincinnati Enquirer, Dec. 2, 197 p. B-1.

3. By law utilities are rewarded for capital investment. Prudently manage utilities earn revenue calculated on the amount of money they invest in plant ar equipment. Ratemaking will be discussed in greater detail in chapter five.

4. Nuclear construction experiences the most severe "regulatory lag," that peri from drawing board until a plant is on line. For coal-fired electric utilities the lag about 40%–50% shorter than for nuclear plants. Office of Technology Assessmer Nuclear Power in an Age of Uncertainty 31, OTA-E-216 (Washington: Februa 1984). By one measurement Zimmer's lag is not complete. The decision to buil nuclear plant was formalized in January 1968. To date, the plant sits idle while t partners are responsible for interest payments on construction loans. The plann conversion will not be complete before 1991.

5. Howard Wilkinson, CG&E Drops Nuclear Plans after Fifteen Years, Cinc nati Enquirer, Jan. 23, 1984, p. A-1.

6. Cincinnati Gas & Electric Annual Report from 1968 at 19.

7. Cincinnati Gas & Electric Co. v. General Electric, Dkt. No. C-1-84-0988 (U D.C. Ohio) (Amended Complaint and Jury Demand, July 10, 1984).

8. Nuclear Regulatory Commission, Order to Show Cause and Order Im diately Suspending Construction, Dkt. No. 50-358, Constr. Permit No. CPPR EA 82-129 (Nov. 12, 1982).

9. Government Accountability Project, Request for an Investigation before

29. Id. at 2–30 to 2–31; and 3–3 to 3–5.
30. Id. at 3–23.
31. Id. at 3–14.
32. Id. at 2–33; and 3–15 to 3–16.
33. Id. at 3–17 to 3–25.
34. Id. at 4–12.
35. Id. at 3–13 to 3–14.
36. Nuclear Regulatory Commission/Office of Inspector and Auditor, Special Inquiry re: Adequacy of IE Investigation at the William H. Zimmer Nuclear Power Station, 50-838/ 80–9 (Aug. 7, 1981) at 2:

> In summary the Region III investigation did not adequately pursue all of the allegations in sufficient depth or breadth and lacked adequate documentation.

37. $3.4 Billion to Convert Ohio Plant, N.Y. Times, Aug. 2, 1984, p. D-1 23, col. 6.
38. Grieves, A $1.6 Billion Nuclear Fiasco, Time, Oct. 31, 1983, p. 96.
39. Office of Consumers' Counsel v. Public Utilities Comm'n, 67 Ohio St. 2d 265, 423 N.E.2d 820 (1981).
40. 423 N.E.2d at 823.
41. Note 21 §1.3.
42. Note 37 at 26.
43. Nuclear Regulatory Commission/Office of Inspection and Enforcement, Report to Congress in Improving Quality and the Assurance of Quality in the Design and Construction of Commercial Nuclear Power Plants (Washington: Nuclear Regulatory Commission March 19, 1984). The report, which used Zimmer as a case study, concluded, at 2–2 to 2–4:

> The principal conclusion of this study is that nuclear construction projects having significant quality-related problems in their design or construction were characterized by the inability or failure of utility management to effectively implement a management system that ensured adequate control over all aspects of the project.
>
> There are two major corollary findings associated with management capability and effectiveness. First, in today's environment, prior nuclear design and construction experience of the collective project team (defined as the architect-engineer (A/E), nuclear steam supply system manufacturer (NSSS), construction manager (CM), and owner), is essential, and inexperience of some members of the project team must be offset and compensated for by experience of other members of the team.
>
> The second corollary finding is that in the past, the NRC has not adequately assessed the factors of management capability and prior nuclear experience in its pre-construction permit reviews and inspections.
>
> The case studies found that while poor craftsmanship played a role in some of the major quality related problems, it was an effect, not the cause, of the underlying problems. The principal underlying cause of poor craftsmanship in constructing nuclear power plants, as well as the quality problem, was found to be poor utility and project management.

44. Governor's Task Force Report on Public Service Company of Indiana Marble Hill Station (Feb. 1984) at 1–10 (hereinafter referred to as Governor's Task Force Report). PSI has excess capacity problems even without the addition of Marble Hill. In Office of Utility Consumer Counsel v. Public Service Co. of Indiana, Utility Law Reports-State ¶24,422 (May 17, 1984), the Indiana Court of Appeals upheld the PSC's order that allowed PSI to include 50.5% of a recently completed coal plant in its rate base. The PSC found that with the addition of the new unit, PSI's reserve capacity rose to 47% and was "estimated to drop to 27% by 1986 due to increased demand," ¶24,422.04. This case is no longer the law in Indiana; see chapter five.
45. Report to Congress, note 43 at 4–9 to 4–10.

46. In re Public Service Company of Indiana, NRC Dkt. No. 50-546 and 50-547, Order Confirming Suspension of Construction (Aug. 15, 1979). Construction Problems at Marble Hill Nuclear Facility: NRC Oversight, Hearings before A Subcommittee of the Committee on Government Operations, House of Representatives, 96th Cong. 1st Sess. (Nov. 27 and 28, 1979).

47. Governor's Task Force Report, note 44, Appendix A at 3.

48. In re Northern Indiana Public Service Co., Dkt. No. 36689 (Aug. 11, 1982).

49. Governor's Task Force, note 44 at 1.

50. Id. at 6.

51. John Bussey, Public Service of Indiana Says It May Be Forced to Seek Protection, Wall Street Journal, Feb. 14, 1984, p. 16.

52. Public Service Indiana Is Charged on Marble Hill Project in $466 Million Suit, Wall Street Journal, Feb. 13, 1984, p. 5.

53. Report to Congress, note 43, Appendix A at A.7 to A.11.

54. Governor's Task Force, note 44 at 13.

55. Note 48.

56. Governor's Task Force Report, note 44 at 13–14.

57. G.E.'s Mark II containment system has been the subject of much controversy and is one of the issues being litigated by CG&E, see note 7.

58. Willmott and Harrison, Report on Shoreham at 15–16 in Report of the New York State Fact Finding Panel on the Shoreham Nuclear Power Facility (Dec. 1983) (hereinafter referred to as Governor's Shoreham Report).

59. To date no utility has gone bankrupt. Briefly, should bankruptcy occur, electricity would not stop; however, the corporate form would be altered resulting in higher management costs.

60. Matthew Wald, LILCO's Workers Reject a Pay Cut and Start Strike, N.Y. Times, July 11, 1984, p. 1, col. 2.

61. Construction costs per kilowatt at Shoreham, for example, at $3,500/kw are four times higher than the Duke Power's McGuire 1 and 2 at $840/kw. See Office of Technology Assessment, Nuclear Power in an Age of Uncertainty at 60, OTA-E-216 (Washington: February 1984). See also Ron Winslow, Atomic Speed: Utility Cuts Red Tape, Builds Nuclear Plant Almost on Schedule, Wall Street Journal, Feb. 22, 1984, p. 1, col. 1.

62. Governor's Shoreham Report, note 58 at 18–28.

63. Ron Winslow, LILCO's Bid to Spread Nuclear Costs Riles Customers and State Officials, Wall Street Journal, Sept. 1, 1983, p. 27.

64. Christopher Falvin, Nuclear Power: The Market Test 39 (Washington: Worldwatch 1983).

65. James Barron, Burden of LILCO Bills Could Slow Up Long Island's Economy, N.Y. Times, Aug. 14, 1983, p. 6-E.

66. Id.

67. Department of Energy/Energy Information Agency, Nuclear Plant Cancellations: Causes, Costs, and Consequences 57 (Washington: Department of Energy 1983); and In re Rochester Gas & Electric Corp., (NYPSC) 45 P.U.R. 4th 386 (1982). The New York PSC is having second thoughts about the percentage of abandonment costs that should be assessed on the ratepayers. Recently, the PSC allowed only 70% of the planning costs for a nuclear reactor owned by LILCO and the New York State Electric and Gas Co. to be passed on because the decision to abort the plant should have been made sooner. See Michael Oreskes, 9.6% Rise in LILCO Rates Approved by the PSC, N.Y. Times, Aug. 16, 1984, p. 1-B 15, col. 1. PSC Opinion No. 84-20 (July 17, 1984). In this opinion, the PSC ordered that a cap of $5.4 billion be placed on costs to be recovered from Nine Mile Point No. 2 generating station.

68. County of Suffolk v. Long Island Lighting Co., 728 F.2d 52 (2d Cir. 1984).

69. Oreskes, note 67.

Comparison of Zimmer, Marble Hill, and Shoreham

	Zimmer	Marble Hill	Shoreham
Originally proposed	1969	1973	1965
Ground broken	1972	1978	1969
Completion date:			
original estimate	1975	1982 for 1st unit 1984 for 2d unit	1975
latest estimate	1986 (before cancellation)	1988 and 1990 (before cancellation)	1986
Percent completed	97%	50% of 1st unit 35% of 2d unit	98%
Original cost estimate	$240 million	$1.4 billion	$265 million
Current cost estimate	$3.1 billion (before cancellation)	$7 billion (before cancellation)	$5+ billion
Amount invested in project now	$1.7 billion	$2.8 billion	$4.1 billion
Generating capacity proposed	810 megawatts	2260 megawatts	820 megawatts

70. See table above.
71. Mark Hertsgaard, Nuclear, Inc. (New York: Pantheon Books, 1983).
72. See table below.

Summary of Relationship of Project Role to Prior Nuclear Experience at Time Quality Problems Occurred

Characteristics	Zimmer	Marble Hill	Shoreham
Design quality problems			X
Construction quality problems	X	X	X
First nuclear project	X	X	
Inexperienced A/E			
Inexperienced construction manager	X	X	X
Inexperienced nuclear constructor	X	X	X
Members of project team inexperienced in role assumed	X	X	X
Inexperience of project team member contributed to quality problem	X	X	X

73. Report to Congress, note 43.
74. Id. at 2-2 to 2-3.

3. *Public Participation*

1. For example, Peter Navarro, The Dimming of America: The Real Costs of Electric Utility Regulatory Failure (Cambridge: Ballinger Publishing Co. 1985).
2. David Roe, Dynamos and Virgins (New York: Random House 1984); Union of Concerned Scientists, Safety Second: A Critical Evaluation of the NRC's First Decade (Washington: Union of Concerned Scientists February 1985); Ian Sanderson, The

Nuclear Controversy: Unequal Competition in Public Policymaking (Milton, England: Open University May 1980); and John Milkovich, Consumers' Challenges to the Construction of Nuclear Power Plants—Closing the Door on Armageddon, 19 New England Law Review 19 (1983–84).

3. Navarro, note 1, and Peter Huber, Safety and the Second Best: The Hazards of Public Risk Management in the Courts, 85 Columbia Law Review 277 (1985). Mr. Huber argues that court review of risk management increases risk.

4. Union of Concerned Scientists, note 2.

5. The text and structure of the Constitution have been read to require notice and an opportunity to be heard when individual rights are affected by public action. Laurence H. Tribe, American Constituional Law §10-1 and §10-7 to -10 (Mineola: The Foundation Press 1978); Richard J. Pierce, Jr., Sidney A. Shapiro, and Paul R. Verkuil, Administrative Law and Process §6.3 (Mineola: The Foundation Press 1985); and Bernard Schwartz, Administrative Law, ch. 5 (Boston: Little, Brown 2d ed. 1984).

6. George T. Mazuzan and J. Samuel Walker, Controlling the Atom: The Beginning of Nuclear Regulation 1946–1962, ch. 6 (Berkeley: University of California Press 1984); Sanderson, note 2; Harold Green, Public Participation in Nuclear Power Plant Licensing: The Great Delusion, 15 William and Mary Law Review 503 (1974).

7. Gerald E. Frug, The Ideology of Bureaucracy in American Law, 97 Harvard Law Review 1276, 1297–1305 (1984).

8. Baruch Fischhoff, Sarah Lichtenstein, Paul Slovic, Stephen L. Derby, and Ralph L. Keeney, Acceptable Risk (Cambridge: Cambridge University Press 1981), and William W. Lowrance, Of Acceptable Risk: Science and the Determination of Safety (Los Altos: William Kaufman 1976).

9. Richard B. Stewart, The Reformation of American Administrative Law, 88 Harvard Law Review 1667 (1975).

10. Mazuzan, note 6, and Green, note 6.

11. K. David Pijawka, The Pattern of Public Response to Nuclear Facilities: An Analysis of the Diablo Canyon Nuclear Generating Station, in Martin J. Pasqualetti and K. David Pijawka, Nuclear Power: Assessing and Managing Hazardous Technology (Boulder: Westview Press 1984).

12. Steven Ebbin, Citizen Groups and the Nuclear Power Controversy 187 (Cambridge: MIT Press 1977).

13. Legal Consequences of Nuclear Accidents and Shutdowns: Transcript of Proceedings Held in Hershey, Pennsylvania 16 (July 27–28, 1979).

14. Harold Green, Licensing Reform in ALI/ABA Atomic Energy Licensing and Regulation 235, 240 (Philadelphia: American Law Institute 1985).

15. John E. Chubb, Interest Groups and the Bureaucracy: The Politics of Energy, ch. 4 (Stanford: Stanford University Press 1983).

16. Ebbin, note 12 at 189.

17. William T. Gormley, Jr., The Politics of Public Utility Regulation (Pittsburgh: University of Pittsburgh Press 1983); Keith Wiens, Citizens Perspective in the Wolf Creek Rate Case, 33 Kansas Law Review 469 (1985) (example of grassroots organization); and Note, the Office of Public Counsel: Institutionalizing Public Interest Representation in State Government, 64 Georgetown Law Journal 895 (1976) (discussion of proxy agencies).

18. James McGrangey Jr., The Slow Return of Reason to Nuclear Regulation, 112 Public Utilities Fortnightly 13 (Sept. 29, 1983), and Brian Moline, Wolf Creek and the Rate Making Process, 33 Kansas Law Review 509 (1985) (examples of project-specific intervention).

19. Union of Concerned Scientists, note 2, and Association of the Bar of the City

of New York, Electricity and the Environment: The Reform of Legal Institutions 136–142 (St. Paul: West Publishing Co. 1972).

20. Pierce et al., note 5 §6.4, and Bernard Schwartz, note 5 at ch. 4.

21. 42 U.S.C. §§2011–2296 (1982).

22. 42 U.S.C. §4331–4335 (1982).

23. 42 U.S.C. §2239(a) (1982).

24. 10 C.F.R. Part 20, App. A (1984).

25. 10 C.F.R. §51 (1984).

26. 10 C.F.R. §2.714 (1984).

27. 10 C.F.R. §2.740 (1984).

28. Donald Stever, Seabrook and the Nuclear Power Commission (Hanover: University Press of New England 1980).

29. 10 C.F.R. §2.762 (1984).

30. Stever, note 28.

31. Roe, note 2 at 48:

> Desultory intervenors were soon defeated by the slow pace of the hearings, the need to wait sometimes for days for the chance to ask for a few pertinent questions of an important witness, and the unpredictability of timing that required almost full-time presence in the hearing room to be sure of catching one's turn. Regulation also seemed to breed complication, a process on which both the regulator and the regulated thrived. Frequently, listening to testimony about one specialized piece of the puzzle or another, I had the sensation that only two people in the world really understood it: the PG&E employee whose domain it was, and the civil servant on the PUC staff who was responsible for reviewing the part of the operation each time a rate case came around.

32. Roe, note 2; Mancur Olson, the Logic of Collective Action: Public Goods and the Theory of Groups, ch. 1 (Cambridge: Harvard University Press 1965), and James M. Buchanan and Gordon Tullock, The Calculus of Consent: Logical Foundations of Constitutional Democracy, ch. 5 (Ann Arbor: University of Michigan Press 1965).

33. Office of Consumers' Counsel v. Public Util. Comm'n, 48 Ohio St. 2d 449, 391 N.E.2d 311 (1979), and Barash v. Pennsylvania Public Util. Comm'n, 490 A.2d 806 (1985).

34. Citizens Action Coalition of Indiana, Inc. v. Northern Indiana Public Service Co., Dkt. No. 1185-470 Slip. op. (Supreme Ct. 1985).

35. Unless otherwise noted, the facts about the Zimmer cancellation were derived from local newspaper accounts in the following newspapers:

Cincinnati Post	July 28, 1982, p. 8A, col. 1.
	Aug. 21, 1982, p. 7A, col. 1.
	Sept. 23, 1982, p. 14C, col. 6
	Oct. 1, 1982, p. 12A, col. 5.
	Oct. 29, 1982, p. 12B, col. 4.
	Feb. 21, 1983, p. 3B, col. 1.
	June 3, 1983, p. 10B, col. 6.
	July 13, 1983, p. 1B, col. 4.
	Oct. 11, 1983, p. 10A, col. 5.
	Oct. 31, 1983, p. 10A, col. 1.
	Jan. 21, 1984, p. 1A, col. 4.
Cincinnati Enquirer	Oct. 27, 1982, p. C-2, col. 1.
	Oct. 29, 1982, p. A-1, col. 3.
	Nov. 13, 1982, p. A-1, col. 1.
	Dec. 13, 1982, p. A-1, col. 1.

	Sept. 21, 1983, p. F-2, col. 1. Sept. 30, 1983, p. A-1, col. 6. Oct. 14, 1983, p. A-14, col. 4. Jan. 22, 1984, p. A-1, col. 5.
New York Times	Sept. 19, 1982, p. L-34, col. 2. Nov. 17, 1982, p. A-16, col. 5. Oct. 8, 1983, p. L-27, col. 4.
Wall Street Journal	Dec. 15, 1982, p. 20, col. 1.

36. Interview with Gretchen Hummel, Legal Counsel, Ohio Consumers' Counsel, March 20, 1984 in Columbus, Ohio.
37. 15 N.R.C. 1549 (June 1982).
38. Mrs. Leigh died in 1977, and her contentions became moot.
39. Telephone interview with Mr. John Woliver, legal counsel for Dr. Fankhauser.
40. Telephone interview with D. David Altman, Chair Environmental Advisory Council, City of Cincinnati.
41. Cover Story, Pulling the Nuclear Plug, 123 Time 34, 36 (Feb. 13, 1984).
42. Telephone interview with W. Peter Heile, Assistant City Solicitor for the City of Cincinnati.
43. 13 N.R.C. 36–41 (Jan. 1981).
44. 11 N.R.C. 511 (April 1980).
45. Office of Nuclear Regulation, Report of the NRC Evaluation Team on the Quality Construction at the Zimmer Nuclear Power Station, NUREG-0969 (Washington: Nuclear Regulatory Commission April 1983).
46. 16 N.R.C. 210 (June 1982).
47. 17 N.R.C. 760 (June 1983).
48. 10 C.F.R. Part 50, App. E, §IV(D)(3) (1984).
49. 15 N.R.C. 1549, 1570 (June 1982).
50. 10 C.F.R. §50.47(a)(1) (1984).
51. 16 N.R.C. 109 (July, 1982).
52. Torrey Pines Technology, Independent Review of Zimmer Project Management GA Project No. 2474 (Aug. 1983).
53. 17 N.R.C. 731 (May 1983).
54. Torrey Pines, note 52.
55. Ohio Revised Code §4905.04 (Page 1977).
56. Ohio Revised Code §4903.12 (Page 1977).
57. Ohio Revised Code §4903.13 (Page 1977).
58. Ohio Revised Code §4905.26 (Page 1983 Supp.).
59. Ohio Revised Code §4903.221 (Page 1983 Supp.).
60. Ohio Revised Code §4903.082 (Page 1983 Supp.).
61. Cleveland Electric Illuminating Co. v. PUC, 173 Ohio St. 450 (1962).
62. Ohio Revised Code §4911 (Page 1977).
63. Ohio Revised Code §4911.15; §4911.03 (Page 1977).
64. Ohio Revised Code §4911.15 (Page 1983 Supp.).
65. Interview with Gretchen Hummel and Janine Migden, OCC (March 20, 1984).
66. This issue is developed more fully in chapters four and five.
67. Consumers' Counsel v. Ohio Public Utilities Comm'n, 58 Ohio St.2d 449 (1979).
68. Id. at 455.
69. 67 Ohio St. 2d 153, 154 (1981).
70. Complaint #84-259-EL-CSS filed March 2, 1984.
71. William A. Spratley, Let's Study Zimmer, Office of Consumers' Counsel.

72. Report of Proceedings of Consumers' Counsel Advisory Committee, I-60–65.
73. Id. at I-99.
74. Complaint #84-259-EL-CSS.
75. Id.
76. Interview with Gretchen Hummel and Janine Migden.
77. Report of Proceedings of Consumers' Counsel Advisory Committee at I-58.
78. Id. at I-87.
79. Frug, note 7; Stewart, note 9; and Douglas Yates, Bureaucratic Democracy: The Search for Democracy and Efficiency in American Government (Cambridge: Harvard University Press 1982).
80. Benjamin R. Barber, Strong Democracy: Participatory Politics for a New Age (Berkeley: University of California Press 1984).
81. Albert O. Hirschman, Exit, Voice, and Loyalty (Cambridge: Harvard University Press 1970).
82. Bruce A. Ackerman, Social Justice in the Liberal State (New Haven: Yale University Press 1980).
83. Gormley, note 17, and Douglas Anderson, Regulatory Politics and Electric Utilities: A Case Study in Political Economy (Boston: Auburn House 1981).
84. Keith Wiens, Citizen Perspective in the Wolf Creek Rate Case, 33 Kansas Law Review 469 (1985), and John M. Simpson, Equal Rights for Citizens before Administrative Agencies, 33 Kansas Law Review 503 (1985).
85. Energy Information Administration, Nuclear Plant Cancellations: Causes, Costs, and Consequences, DOE/EIA-0392 (Washington: Department of Energy April 1983).
86. Woliver, note 39.
87. Guido Calabresi, A Common Law for the Age of Statutes (Cambridge: Harvard University Press 1982).
88. Bruce A. Ackerman and William Hassler, Clean Coal/Dirty Air (New Haven: Yale University Press 1981).
89. Industrial Union Department v. American Petroleum Institute, 448 U.S. 607 (1980) (Rehnquist, J. concurring), and American Textile Manfacturers Institute, Inc. v. Donovan, 452 U.S. 490 (1981) (Rehnquist, J. dissenting).

4. Nuclear Power Market

1. James Cook, Nuclear Follies, Forbes 82 (Feb. 11, 1985).
2. The remainder is generated by federal, state, local, and cooperative utilities. William F. Fox, Jr., Federal Regulation of Energy §30.01 (Colorado Springs: Shepard's/McGraw Hill 1983). See also Paul L. Joskow and Richard Schmalensee, Markets for Power: An Analysis of Electric Utility Deregulation at 12, tables 2.1 and 2.2 (Cambridge: MIT Press 1985).
3. 42 U.S.C. §§8301-8484 (1982).
4. The existence of the conservation option has had a significant impact on the market for the delivery of electricity by increasing competition. The point will be developed later in the chapter.

Conservation, whether defined as a reduction in demand or an increase in energy efficiency (for example, through the use of insulation), can be treated as a resource. Douglas Norland and James Wolf, Utility Conservation Programs: A Regulatory and Design Framework, 116 Public Utilities Fortnightly 15 (July 25, 1985); Geoffrey C. Crandell, Devere L. Elgas, and Martin G. Kushler, Making Residential Conservation Service Work: A Trilogy of Perspectives, 115 Public Utilities Fortnightly 28 (Jan. 10, 1985); and David Roe, Dynamos and Virgins (New York: Random House 1984).

5. General Accounting Office, Report to the Chairman, Subcommittee on Energy Conservation and Power, House Committee on Energy and Commerce, Analy-

sis of the Financial Health of the Electric Utility Industry, GAO/RCED-84-22 (June 11, 1984), and Electric Power: Potential for Shortages in the 1990's, Hearings before the Committee on Energy and Natural Resources United States Senate 99th Cong. 1st Sess. (July 23 and 25, 1985), testimony of William Hogan and testimony of Dr. Mark French.

6. Armen A. Alchian and William R. Allen, Exchange and Production: Competition, Coordination, and Control (Belmont, Cal.: Wadsworth Pub. Co. 3d ed. 1983).

7. Stephen Breyer, Regulation and Its Reform, ch. 1 (Cambridge: Harvard University Press 1982). Judge Breyer lists several examples of market failure, including: monopoly, inadequate information, spillovers, economic rents, excessive competition, unequal bargaining power, moral hazard, and scarcity.

8. Breyer, note 7 at 33. "Occasionally governmental intervention is justified on the ground that, without it, firms in an industry would remain too small or would lack sufficient organizations to produce their product efficiently."

9. An externality can be defined as a cost (or benefit) not recognized in the unregulated price of a good or service. Breyer, note 7 at 23. With nuclear power, many of the safety problems, including the costs of lowering probabilities of high-risk events, are externalities. They impose costs that would be placed on society but for regulation.

10. Richard L. Gordon, Reforming the Regulation of Electric Utilities, ch. 1 (Lexington: Lexington Books 1982); Joskow and Schmalensee, note 2 ch. 3; Keith M. Howe and Eugene F. Rasmussen, Public Utility Economics and Finance, ch. 2 (Englewood Cliffs: Prentice-Hall 1982); and 1 Alfred E. Kahn, The Economics of Regulation: Principles and Institutions, ch. 1 (New York: John Wiley and Sons 1970). See Kenneth D. Boyer, Testing the Applicability of the Natural Monopoly Concept, in Werner Sichel and Thomas G. Gies (eds.), Applications of Economic Principles in Public Utility Industries 1 (Ann Arbor: University of Michigan 1981).

11. Breyer, note 7 at 15–16.

12. John N. Drobak, From Turnpike to Nuclear Power: The Constitutional Limits on Utility Rate Regulation, 65 Boston University Law Review 65 (1985), and Charles F. Phillips, Jr., The Regulation of Public Utilities, ch. 1 (Arlington: Public Utilities Report, Inc. 1984).

13. Kahn, note 10.

14. Leonard S. Hyman, America's Electric Utilities: Past, Present, and Future 199–200 (Arlington: Public Utilities Reports, Inc. 2d ed. 1985), and Technology Futures, Inc. and Scientific Foresight, Inc., Principles for Electric Power Policy, ch. 5 (Westport: Quorum Books 1984).

15. Phillips, note 12 at 41.

16. Hyman, note 14 chs. 13 and 14. Large central power stations are facing more competition. John S. Ferguson, Is Central Power Station Generation Becoming a White Elephant? 115 Public Utilities Fortnightly 32 (March 21, 1985); Victor A. Canto and Charles W. Kaldec, The Shape of Energy Markets to Come, 117 Public Utilities Fortnightly 21 (Jan. 9, 1986); and Charles F. Phillips, Jr., The Changing Structure of the Public Utility Sector, 117 Public Utilities Fortnightly 13 (Jan. 9, 1986).

17. Howe and Rasmussen, note 10 at 20: "Empirical studies indicate that price elasticity of demand for utility services is relatively inelastic"; and Joskow and Schmalensee, note 10 at 156: "Numerous studies of the demand for electricity indicate that the short-run elasticity of demand at current (1983) prices is much less than 1 and that long-run price elasticity is around 1"; and Hyman, note 14 at 43, table 6-2.

18. Roger D. Colton, Excess Capacity: Who Gets the Charge from the Power Plant? 34 Hastings Law Journal 1133 (1983); Dialogue, Excess Capacity, 35 Hastings Law Journal 721 (1984); Excess Capacity: A Case Study in Ratemaking Theory and

Application, 20 Tulsa Law Journal 402 (1985); and Alvin Kaufman, Kevin Kelly, and Ross Hemphill, Commission Treatment of Overcapacity in the Electric Power Industry (Columbus: National Regulatory Research Institute 1984).

19. Katherine A. Miller, Strategies for an Electric Utility Industry in Transition, 116 Public Utilities Fortnightly 27 (June 13, 1985).

20. Technology Futures, Inc. and Scientific Foresight, Inc., note 14 at 122.

21. Regarding the nuclear future, see Ronald Kluch, A Second Nuclear Era? 116 Public Utilities Fortnightly 15 (Oct. 31, 1985), and John O. Sillin, Managing to Reduce Nuclear Financial Risks, 114 Public Utilities Fortnightly 26 (Oct. 11, 1984).

22. Although estimates vary, most studies conclude that coal plants are cheaper to build than nuclear plants. Charles Komanoff, Assessing the High Costs of New U.S. Nuclear Power Plants (June 1984) (paper on file with author), and Office of Technology Assessment, Nuclear Power in an Age of Uncertainty 64-66 (Washington, D.C.: U.S. Congress, Office of Technology Assessment, OTA-E-216 Feb. 1984). For a contrary view, see Lewis J. Perl, Estimated Costs of Coal and Nuclear Generation (Dec. 12, 1978) (paper on file with author). Another version is that there are regional differences between coal and nuclear. Energy Information Administration, Projected Costs of Electricity from Nuclear and Coal-Fired Power Plants, DOE/EIA-035611 (Washington: Department of Energy August 1982).

23. Peter Navarro, The Dimming of America: The Real Costs of Electric Utility Regulatory Failure (Cambridge: Ballinger Publishing Co. 1985); Edison Electric Institute, Report of the Edison Electric Institute on Nuclear Power (Feb. 1985); Electric Power: Potential for Shortages in the 1990's, note 5; and Edward J. Mitchell and Peter R. Chaffetz, Toward Economy in Electric Power (Washington: American Enterprise Institute 1975).

24. Technology, Futures, Inc. and Scientific Foresight, Inc., note 14 ch. 2 (listing six electric scenarios: Average Future, Nuclear Resurgence, Mega-plant, Small Coal, Post-industrial, and Economic Malaise).

25. General Accounting Office, note 5.

26. Leonard Hyman, Heidimarie West, Susan Malley, and Victor Borum, Evidence on the Effect of Nuclear Construction on Electric Utility Share Prices (submitted for publication in Financial Management 1985) (copy on file with author), and Office of Technology Assessment, note 22 at 46–52.

27. Carl H. Hobelman, George M. Knapp, and Kevin M. Walsh, Construction Work in Progress for Electric Utilities: A Compendium of Comments Presented to the Federal Energy Regulatory Commission in Docket No. RM81-38 (7/27/81), in Donald R. Allen (ed.), Electric Power: Current Issues in Regulation and Finance 89 (New York: Practicing Law Institute 1982).

28. Id. at 92.

29. Navarro, note 23 ch. 2.

30. Howe and Rasmussen, note 10 at 238.

31. Amory Lovins, Soft Energy Paths: Toward a Durable Peace §2.4 (Cambridge: Ballinger Publishing Co. 1977). Logically, steadily rising prices will eventually see utilities pricing themselves out of the market. For a nonregulated firm, that means bankruptcy. Although threatened, utility bankruptcy is unlikely. Evan D. Flashchen and Michael J. Reilly, Bankruptcy Analysis of a Financially-Troubled Electric Utility, 22 Houston Law Review 965 (1985).

32. The discussion of leveraging and financing relies on Hyman, note 14 and Howe and Rasmussen, note 10.

33. Although competition is clearly increasing in the electric industry, total deregulation in the near term is unlikely. Primarily, some segments of the electricity fuel cycle are more competitive than others. Breaking the fuel cycle into generation, transmission, and distribution, and market for generation is more competitive than the transmission and distribution (or retail sales) levels. The market is complicated by

the fact that most electric utilities integrate those three functions. Joskow and Schmalensee, note 2; William J. Collins, Electric Utility Rate Regulation: Curing Economic Shortcomings through Competition, 19 Tulsa Law Journal 141 (1983); Leonard S. Hyman and Ernst R. Habicht, Jr., State Electric Utility Regulation: Financial Issues, Influences, and Trends (to be published in Annual Energy Review (1986) (copy on file with author); Gary D. Allison, Imprudent Nuclear Power Construction Projects: The Malaise of Traditional Public Utility Policies, 13 Hofstra Law Review 507 (1986); Essay, Efficiency, and Competition in the Electric Power Industry, 88 Yale Law Journal 1511 (1979); and note 16.

34. Joseph P. Tomain, Energy Law, ch. 5 (St. Paul: West Publishing Co. 1981); Ernest Gellhorn and Richard J. Pierce, Regulated Industries, ch. 4 (St. Paul: West Publishing Co. 1982); and Phillips, note 12 at ch. 5.

35. Drobak, note 12, and Allison, note 33.

36. Harvey Averch and Leland L. Johnson, Behavior of the Firm under Regulatory Constraint, 52 American Economic Review 1052 (1962), and Kahn, note 10, vol. 2 chs. 2 and 3.

37. Navarro, note 19 at 15–16, argues that the electric utility industry is experiencing a "reverse AJ effect," by which he means that the industry is "witnessing a dramatic *underinvestment* in new power plants and conversion and conservation projects." Navarro attributes this "underinvestment" to PUC behavior.

38. 1 A. J. G. Priest, Principles of Public Utility Regulation: Theory and Application 191 (Charlottesville: Michie Co. 1969), and Phillips, note 12 at 338.

39. Designating what properties should be included or excluded from the rate base is highly controversial, as will be discussed in chapter five. Suffice it to say here that the traditional "used and useful" standard has been hard to apply. Melvin G. Dakin, The Changing Nature of Public Utility Regulation: The Used and Useful Property Rate Base versus the Capitalization Rate Base in the Nuclear Age, 45 Louisiana Law Review 1033 (1985).

40. Phillips, note 12 at ch. 9.

41. Illinois Office of Consumer Services, A Consumer's Guide to the Economics of Electric Utility Ratemaking (1980).

42. Benhamin Graham, David Dodd, and Sidney Cottle, Security Analysis: Principles and Techniques 268 (New York: McGraw-Hill 4th ed. 1962).

43. Id. at 271.

44. Hobelman et al., note 27 at 74.

45. Bluefield Co. v. Public Serv. Comm'n, 262 U.S. 679, 692–93 (1923).

46. Federal Power Comm'n v. Hope Natural Gas Co., 320 U.S. 591, 633 (1944).

47. Energy Information Administration, Nuclear Plant Cancellations: Causes, Costs, and Consequences 19, DOE/EIA-0392 (Washington: Department of Energy April 1983).

48. DOE Seeks Investors Shielded from 70% of Nuclear Unit Cancellation Costs, Electric Utility Week (May 30, 1983).

49. Maureen A. Crawfis, A New Approach to Allocating Financial Responsibility for Cancelled Nuclear Units: Consumers' Counsel v. Public Utility Comm'n, 13 University of Toledo Law Review 1469, 1475 (1982).

50. Richard Pierce, The Regulatory Treatment of Mistakes in Retrospect: Canceled Plants and Excess Capacity, 132 University of Pennsylvania Law Review 497, 519 (1984).

51. Electric Utility Week, note 48.

52. Navarro, note 19 chs. 3–5, and Andrew S. Carron and Paul W. MacAvoy, The Decline in Service in the Regulated Industries (Washington: American Enterprise Institute 1981).

53. Re Wolf Creek Nuclear Generating Facility, Docket Nos. 120,924-U and 142,098-U (KG&E); 142,099-U (KCP&L): 142,100-U (Sept. 27, 1985); Duncan

Wyse, Avoided Costs Ratemaking for Diablo Canyon, Application No. 84-06-014 (Sept. 10, 1985) (copy on file with author); prepared testimony of Ron Knecht, California Public Utilities Commission, Order Instituting Investigation No. 85-05-001 Southern California Edison Company-Respondent, Ratemaking for the Palo Verde Nuclear Project (June 17, 1985) (copy on file with author); Tomain, note 37; A. Lawrence Kolbe, How Can Regulated Rates and Companies Survive Competition, 116 Public Utilities Fortnightly 25 (April 4, 1985); and Kermit R. Kubitz, The Energy Utilities: How to Increase Rewards to Match Increasing Risks, in Jonathan David Aronson and Peter F. Cowhey (eds.), Profit and the Pursuit of Energy: Markets and Regulation (Boulder: Westview Press 1983).

54. Joskow and Schmalensee, note 2; Collins, note 33; Andrew Varley, Is the Electric Utility Industry Ready for Deregulation? 116 Public Utilities Fortnightly (Sept. 19, 1985); and Leonard Hyman, The Future of the Electric Utility Industry (June 24, 1985) (copy of paper on file with author).

55. John C. Sawhill and Lester P. Silverman, Your Local Utility Will Never Be the Same, Wall Street Journal, Jan. 2, 1986, p. 10, col. 3, and Joseph P. Tomain, Utilities: A Change Is Needed, Cincinnati Enquirer, Dec. 27, 1985.

56. In re Three Mile Island Alert, Inc., 771 F.2d 720 (3rd Cir. 1985).

57. Cuomo v. United States Nuclear Regulatory Comm'n, 772 F.2d 972 (D.C. Cir. 1985).

58. Pierre Tanguy,. Safety and Nuclear Power Plant Standardization: The French Experience, 116 Public Utilities Fortnightly 20 (Oct. 31, 1985).

59. Sillin, note 21.

5. *Accommodating Nuclear Power*

1. John N. Drobak, From Turnpike to Nuclear Power: The Constitutional Limits on Utility Rate Regulation, 65 Boston University Law Review 65 (1985), and Federal Power Commission v. Hope Natural Gas Co., 320 U.S. 59 (1944).

2. PUCs and courts have addressed the problem for various utility functions, such as gas manufacturing plants, pollution equipment, streetcar barns, and steam plants. The abandonment problem is not new; the various rules that will be explained in this chapter are simply given a new application.

3. Aside from their large cost, nuclear plants, it can be argued, have qualitatively different attributes from those of other technologies. From a safety standpoint, nuclear technology contains low-probability, high-risk potentialities. Both safety and finances present transgenerational concerns. Further, transgenerational issues combine to disempower decision makers by distancing the decision from its consequences, thus making nuclear policy making qualitatively different from short-term, less technologically complex policy making. Hans Jonas, The Imperative of Responsibility: In Search of an Ethics for the Technological Age (Chicago: University of Chicago Press 1984), and Robert E. Goodin, Political Theory and Public Policy (Chicago: University of Chicago Press 1982).

4. Specific examples of assumptions that no longer command allegiance are the concept of natural monopoly and the relatively high price elasticity of demand for electricity; see chapter four.

5. Thomas S. Kuhn, The Structure of Scientific Revolutions (Chicago: University of Chicago Press 2d ed. 1970).

6. Amory Lovins, Soft Energy Paths: Toward a Durable Peace (Cambridge: Ballinger Publishing Co. 1977).

7. Id., and James S. Fishkin, The Limits of Obligation (New Haven: Yale University Press 1981).

8. Lester C. Thurow, The Zero–Sum Society: Distribution and the Possibilities for Economic Change (New York: Basic Books 1980).

9. Richard J. Pierce, Jr., The Regulatory Treatment of Mistakes in Retrospect:

Canceled Plants and Excess Capacity, 132 University of Pennsylvania Law Review 497 (1984); Melvin G. Dakin, The Changing Nature of Public Utility Regulation: The Used and Useful Property Rate Base versus the Capitalization Rate Base in the Nuclear Age, 45 Louisiana Law Review 1033 (1985); and Roger D. Colton, Excess Capacity: Who Gets the Charge from the Power Plant? 34 Hastings Law Journal 1133 (1983).

10. Iowa-Illinois Gas & Electric Co. v. Iowa State Commerce Comm'n, 347 N.W.2d 423 (Iowa, 1984), and Roger D. Colton, Excess Capacity: A Case Study in Ratemaking Theory and Application, 20 Tulsa Law Journal 402 (1985).

11. The discussion of how legal rules are manipulated to justify outcomes occupies a large amount of jurisprudential writing. The two major philosophical movements that explore this issue are Legal Realism and Critical Legal Studies. These two movements differ on the degree to which they ascribe indeterminacy to the rule of law. I adopt a traditional liberal approach to law, which accepts a certain amount of indeterminacy in a rule, and believe that legal processes can be established that are fair and from which politically legitimate consensual decisions can emerge. The fact that a choice of tests exists that generate conflicting outcomes is a clear signal that either the policy behind the rule is fragmented or the decision-making process has broken down. In the case of nuclear power, both weaknesses are present.

12. 461 U.S. 190 (1983).

13. Bruce A. Ackerman, Private Property and the Constitution (New Haven: Yale University Press 1977), and Richard A. Epstein, Takings: Private Property and the Power of Eminent Domain (Cambridge: Harvard University Press 1985).

14. Drobak, note 1.

15. William T. Gormley, Jr., The Politics of Public Utility Regulation (Pittsburgh: University of Pittsburgh Press 1983).

16. Robin Lee Fenton, An Example of Regulatory Failure: Why Nuclear Power Plant Construction Costs Should Be Included in the Rate Base during the Construction Period (December 1985) (copy of paper on file with author), and Constance A. Olson, Statutes Prohibiting Cost Recovery for Cancelled Nuclear Power Plants: Constitutional? Pro-Consumer? 28 Journal of Urban and Contemporary Law 345 (1985). Both authors argue that consumers are worse off if CWIP is disallowed. This position is true if and when a plant with a long lead time comes on line, because carrying costs are borne by ratepayers.

17. Schippen Howe, A Survey of Regulatory Treatment of Plant Cancellation Costs, 111 Public Utilities Fortnightly 52 (March 31, 1985); Pierce, note 9, and Energy Information Agency, Nuclear Plant Cancellations: Causes, Costs, and Consequences, DOE/EIA-0392 (Washington: Department of Energy DOE/EIA—April 1983).

18. Harvey Averch and Leland L. Johnson, Behavior of the Firm under Regulatory Constraint, 52 American Economic Review 1052 (1962).

19. Office of Consumers' Counsel v. Public Util. Comm'n, 48 Ohio St. 2d 449, 391 N.E.2d 311 (1979).

20. Ohio Rev. Code Ann. §4909.15 (A)(4) (Page 1981).

21. 67 Ohio St.2d at 163, 423 N.E.2d at 426–27.

22. I have qualified the last two sentences with the word *generally* because taxpayers also absorb some losses.

23. Dayton Power & Light Co. v. Pub. Util. Comm'n, 4 Ohio St.3d 91, 103, 447 N.E.2d 733, 743 (1983). Ohio has continued to disallow costs associated with failed nuclear projects to be passed through to taxpayers. In Columbus & Southern Ohio Elec. Co. v. Pub. Util. Comm'n, 10 Ohio St.2d 12, 460 N.E.2d 1108 (1984), the reviewing court upheld a CWIP disallowance regarding the canceled Zimmer nuclear plant, and in Toledo Edison Co. v. Pub. Util. Comm'n, 12 Ohio St.3d 143, 465 N.E.2d 886 (1984), operating expenses for a canceled unit were disallowed.

24. Id. at 889.
25. Ohio Consumers' Counsel v. Pub. Util. Comm'n, 4 Ohio St.3d 111, 115 (1983).
26. Lower Valley Power & Light, Inc., Dkt. Nos. 9617 sub 11, 9628 sub 6, and 9454 sub 17 (Pub. Serv. Comm'n Wyo., Dec. 2, 1982).
27. Wyo. Stat. §37-2-119 (1977).
28. Pacific Power & Light Co. v. Public Serv. Comm'n, 677 P.2d 799 (Wyo. 1984), cert. denied, 105 S.Ct. 720 (1985).
29. Id. at 806.
30. Or. Rev. Stat. §757-355 (1979).
31. In the Matter of the Anticipated Abandonment of Electric Generating Plant Projects by Portland General Electric Co. & Pacific Power & Light Co., CCH Utilities Law Reporter-State ¶24,202, (Dec. 27, 1983). See also In re Portland General Electric, 49 P.U.R. 4th 274 (Or. PUC, 1982), where PGE's write-off of its $132 million interest in Pebble Springs was absorbed by the utility, not the ratepayers.
32. Re Pacific Power & Light Co., 42 P.U.R. 4th 24 (Mont. PSC, April 18, 1983).
33. Mont. Code Ann. §69-3-109 (1983).
34. See note 32.
35. Id.
36. Re Arizona Pub. Serv. Co., 38 P.U.R. 4th 547, 556 (Ariz. Corp. Comm'n, May 29, 1981).
37. Id.
38. N.H. Rev. Stat. Ann §378:30-a (Supp. 1981).
39. Appeal of Public Service Co. of New Hampshire, 480 A.2d 20 (N.H. 1984).
40. Missouri ex rel. Union Elec. Co. v. Missouri Pub. Serv. Comm'n, 687 S.W.2d 162 (Mo. 1985).
41. Citizens Action Coalition of Indiana, Inc. v. Northern Indiana Public Service Co., Dkt. No. 1185-470 Slip. op. at 6–7 (Supreme Ct. 1985) affirming 472 N.E.2d 938 (Ind. Ct. App. 1984).
42. Pennsylvania Electric Co. v. Pennsylvania Public Util. Comm'n, Slip. op. (Sup. Ct. Dec. 9, 1985). See also Barasch v. Pennsylvania Public Util. Comm'n, 490 A.2d 806 (1985), and Barasch v. Public Util. Comm'n, 491 A.2d 904 (1985).
43. In re Pacific Gas & Electric Co., 34 P.U.R. 4th 1 (Cal. PUC 1979).
44. Re Rochester Gas & Elec. Corp., 45 P.U.R. 4th 386, 388 (N.Y. PSC, Jan. 13, 1982).
45. Shimon Awerbach and Donna Freireich, Nuclear Plant Cancellations: Economic and Legal Bases for Allocating Losses in Michigan State University Public Utility Papers, Award Papers in Public Utility Economics and Regulation (Lansing: Michigan State University 1982).
46. Long Island Lighting Co.—Phase II—Proceeding on Motion of Commission to Investigate Cost of Construction of Shoreham Nuclear Generating Facility, CCH Utilities Law Reporter—State ¶24,922 (Dec. 16, 1985); Matthew Wald, (1985) Lilco Is Blamed for $1.2 Billion of Shoreham Overrun, New York Times, March 14, 1985, p. B2.
47. Re United Illuminating Co., 55 P.U.R. 4th 252 (Conn. Dept. Publ. Util. Control, Aug. 22, 1983). See also Re United Illuminating Co., 62 P.U.R. 4th 319 (Conn. DPUC 1984).
48. Re Boston Edison Co., 46 P.U.R. 4th 431, 461 (Mass. DPU, April 30, 1982).
49. See also Boston Edison Co. v. Department of Public Utilities, 375 N.E.2d 305 (Mass. 1984), in which replacement power costs associated with a shutdown of Pilgrim I were disallowed after a finding that an imprudent management decision was responsible for the delay in reopening.
50. 46 P.U.R. 4th at 459.

51. Attorney General v. Department of Pub. Util., 455 N.E.2d 414, 426 (Mass. 1983).

52. Id. at 424–25.

53. Id. at 479.

54. Id. at 480.

55. 455 N.E.2d at 428 (Liacos, J. dissenting).

56. Central Maine Power Co. v. Maine Publ. Util. Comm'n, 433 A.2d 331 (Me. 1981). See also Re Bangor Hydro-Electric Co., 46 P.U.R. 4th 503 (Me. PUC, April 8, 1982).

57. 46 P.U.R. 4th at 557.

58. Id.

59. Re Union Elec. Co., 53 P.U.R. 4th 565, 592 (Ill. Commerce Comm'n, May 25, 1982).

60. Re Carolina Power & Light Co., 49 P.U.R. 4th 188, 217 (N.C. Util. Comm'n, Sept. 24, 1982).

61. Id. at 218.

62. Re Carolina Power & Light Co., 55 P.U.R. 4th 582, 601 (N.C. Util. Comm'n, Sept. 19, 1983).

63. Id. at 600.

64. Department of Energy/Energy Information Agency, Nuclear Plant Cancellations: Causes, Costs, and Consequences 71, DOE/EIA-0392 (Washington: Department of Energy April 1983):

> A present-value analysis of the costs allocated to the three major payer groups for a hypothetical plant cancellation involving amortization over 10 years, with no return earned on the unamortized balance yielded the following appropriate distribution of costs; utility investors, 30 percent; utility ratepayers, 30 percent; and income taxpayers, 40 percent.

65. North Carolina Utilities Comm'n v. Conservation Council, 320 S.E.2d 679 (N.C. 1984).

66. Washington Util. & Transp. Comm'n v. Pacific Power & Light Co., 51 P.U.R. 4th 158 (Wash. Util. & Transp. Comm'n, Feb. 1, 1983).

67. Id. at 168.

68. Washington Util. & Transp. Comm'n v. Puget Sound Power & Light Co., 54 P.U.R. 4th 480 (Wash. Util. & Transp. Comm'n, July 22, 1983).

69. People's Organization for Washington Energy Resources v. Washington Utilities and Transportation Comm'n, 679 P.2d 922 (Wash. 1984).

70. Re Virginia Electric & Power Co., 54 P.U.R. 4th 46, 49 (Va. State Corp. Comm'n, Aug. 24, 1981).

71. Re Virginia Electric & Power Co., 44 P.U.R. 4th 46, 49 (Va. State Corp. Comm'n, Aug. 24, 1981).

72. Re Atlantic City Electric Co., 51 P.U.R. 4th 109, 115 (N.J. Bd. Pub. Util., Jan. 13, 1983). See also Re Jersey Central Power & Light Co., 428 A.2d 598 (N.J. 1981).

73. In the Matter of the Application of the Washington Water Power Co., Dkt. No. U-1008-219, CCH Utilities Law Reports—State ¶24,271 (Jan. 30, 1985).

74. Re Central Vermont Pub. Serv. Corp., 49 P.U.R. 4th 372, 392 (Vt. Pub. Serv. Bd., Sept. 16, 1982).

75. Id.

76. Jackson v. Metropolitan Edison Co., 419 U.S. 345, 357 (1974).

77. Like rules of law, policies go through recognizable periods. Once a rule or policy has been announced, its second stage begins, during which cases are included within and without the rule, distinctions become fuzzy and sometimes contradictory, and reasoning is questionable, all of which leads to the demise and replacement or

reformulation of the rule or policy. The hallmark of this phase of rule or policy development is a certain amount of lumpiness. Some lumpiness (more charitably, flexibility) is healthy, because the dynamic of the process allows law to adapt. Too much flexibility is properly characterized as confusion, and that is an unhealthy state in which the dynamic breaks down. Practical as well as theoretical problems accompany a breakdown in analogical reasoning. Courts are unclear about the basis for decisions. Judges are given too much leeway in deciding whether to grant or deny relief and base their decisions on unarticulated premises or meta-rules that are more result-oriented than is comfortable. There is a dislocation between the espoused rule and the operative principles. Labels replace logic; prediction is weakened; claims are chilled because transaction costs are too high; and lawyers cannot advise clients as precisely as they might otherwise, because the signals about the likelihood of recovery are unclear. Joseph P. Tomain, Contract Compensation in Nonmarket Transactions, 46 University of Pittsburgh Law Review 867, 890–91 (1985).

78. William F. Fox, Jr., Federal Regulation of Energy, Part Four (Colorado Springs: Shepard's/McGraw-Hill 1983).

79. Id. at §2.02.

80. Federal Energy Reorganization Act, Pub. L. No. 93-428, 88 Stat. 1233 (1974).

81. Department of Energy Organization Act, 42 U.S.C. §§7101-7375 (1982).

82. 42 U.S.C. § 7135(2) (1982).

83. 42 U.S.C. § 7133 (1982).

84. For example, see note 64; Energy Information Administration, Commercial Nuclear Power 1984: Prospects for the United States and the World, DOE/EIA-0438(84) (Washington: Department of Energy November 1984); Energy Information Administration, Survey of Nuclear Power Plant Construction Costs 1984, DOE/EIA-0439(84) (Washington: Department of Energy November 1984); and Energy Information Administration, U.S. Commercial Nuclear Power: Historical Perspective, Current States, and Outlook, DOE/EIA-0315 (Washington: Department of Energy March 1982).

85. Robert Stobaugh and Daniel Yergin (eds.), Energy Future: Report of the Energy Project at Harvard Business School (New York: Random House 1979).

86. Exec. Order No. 11,695, 38 Fed. Reg. 1473 (Jan. 12, 1983); and 38 Fed. Reg. 22,536 (Aug. 22, 1983) and William F. Fox, Jr., Federal Regulation of Energy §§6.03-04 (Colorado Springs: Shepard's/McGraw-Hill, 1983).

87. 15 U.S.C §§751 et seq. (1982).

88. 42 U.S.C. §7151 (1982).

89. The Energy Supply and Environmental Coordination Act of 1974, Pub. L. No. 93-319, 88 Stat. 246 (1974), administered by the FEA, was to encourage utilities to switch to coal. The ESECA was replaced with the stronger Powerplant and Industrial Fuel Use Act, 42 U.S.C. §§8301-8484 (1982), which empowered the DOE to order fuel switching.

90. DOE News Release, May 8, 1984.

91. Id.

92. Compare Energy Information Administration, Commercial Nuclear Power 1984: Prospects for the United States and the World, DOE/EIA-0438 (Washington: Department of Energy November 1984) with Edison Electric Institute, Report of the Edison Electric Institute on Nuclear Power (Washington, D.C.: Edison Electric Institute February 1985).

93. For example, Department of Energy, The National Energy Policy Plan, DOE/S-0014/1 (Washington: Department of Energy October 1983).

94. 42 U.S.C. §§7171-77 (1982).

95. 16 U.S.C §79(a) et seq. (1982).

96. 16 U.S.C. §824(a) (1982).

97. Peter Navarro, The Dimming of America: The Real Costs of Electric Utility

Regulatory Failure (Cambridge: Ballinger Publishing Co. 1985), and William J. Collins, Electric Utility Rate Regulation: Curing Economic Shortcomings through Competition, 19 Tulsa Law Journal 141 (1983).

98. Pub. L. No. 95-615, 92 Stat. 3317 (1978).

99. Fox, note 78 §30.02.

100. 456 U.S. 742 (1982).

101. Navarro, note 97.

102. CCH Util. Law Rept.-Federal ¶13,005 (April 11, 1985).

103. Id. The wisdom (and correctness) of FERC's analysis in this case is questionable. NEPCO was absolved of any imprudence attributable to Boston Edison Co. on the Pilgrim II nuclear project on which they were partners. However, NEPCO's contract with Boston Edison virtually insulated Boston Edison from suit by NEPCO. Thus, the legal characterization of the relationship between NEPCO and Boston Edison becomes crucial. FERC found that Boston Edison's imprudence could not be imputed to NEPCO.

104. Jersey Central Power & Light Co. v. FERC (D.C. Cir. 1984) and NEPCO Municipal Rate Committee v. FERC, 668 F.2d 1327 (D.C. Cir. 1981), cert. denied, 457 U.S. 1117 (1982). See also Union Elec. Co. v. FERC, 6689 F.2d 389 (8th Cir. 1981).

105. 668 F.2d at 1333.

106. Robert Taylor and Matt Moffett, Middle South Utilities Ordered to Share the Costs of Grand Gulf Nuclear Plant, Wall Street Journal, June 14, 1985, p. 4, col. 3.

107. South Dakota Public Utilities Commission v. FERC, 690 F.2d 674 (8th Cir. 1982).

108. 18 C.F.R. §35.26 (1984); Mid-Tex Elec. Co-op, Inc. v. FERC, 773 F.2d 327 (D.C. Cir. 1985); and 48 Federal Register 24,323–24,358 (June 1, 1983).

109. See, e.g., Towns of Concord, Norwood and Wellesley v. FERC, 729 F.2d 824 (D.C. Cir. 1984).

110. 17 FERC (CCH) ¶51, 118 (Nov. 5, 1981).

111. 42 U.S.C. §§10101–10226 (West 1983).

112. Report of the President's Commission on the Accident at Three Mile Island, The Need for Change: The Legacy of TMI 51–56, 61–67 (Washington: October 1979) (Kemeny Commission Report).

113. Mitchell Rogovin, Three Mile Island: A Report to the Commissioners (Washington 1979).

114. The discussion of emergency planning is taken from Donald P. Irwin, State and Federal Roles in Emergency Planning (Sept. 21, 1984). This paper was delivered at the American Law Institute-American Bar Association seminar on Atomic Energy Licensing and Regulation, October 1984. A copy is on file with the author.

115. NUREG-0396, Planning Basis for the Development of State and Local Government Radiological Emergency Response Plans in Support of Light Water Nuclear Power Plants (Washington: Nuclear Regulatory Commission December 1978).

116. NUREG-0654, Criteria for Preparation and Evaluation of Radiological Emergency Response Plans and Preparedness in Support of Nuclear Power Plants (Washington: Nuclear Regulatory Commission January 1980).

117. The rules are contained in 10 C.F.R. §§50.33, 50.47 and 50.54 and Part 50 Appendix E (1984).

118. Irwin, note 114.

119. General Accounting Office, Further Actions Needed to Improve Emergency Preparedness around Nuclear Plants, GAO/RCED-84-83 (Aug. 1, 1984).

120. 735 F.2d 1437 (D.C. Cir. 1984).

121. 753 F.2d 1144 (D.C. Cir. 1985).

122. Nuclear Regulatory Commission Regulatory Reform Task Force, Draft Report SECY-82-447 (Nov. 1982). See also Harold Green, Licensing Reform in American Law Institute/American Bar Association, Atomic Energy Licensing and Regulation (Philadelphia: American Law Institute 1985).

123. Id.

124. American Textile Manufacturers Institute, Inc. v. Donovan, 452 U.S. 490 (1981).

6. *Responsibility and Liability*

1. The critical literature on cost-benefit analysis does not fault the method as much as it faults its overuse by decision makers and their overreliance on it. Economic analysis of legal decisions and, a fortiori, cost-benefit analysis are powerful tools for gathering information. They should not be relied upon as exclusive decision-making tools. This dichotomy is a replay of the separatist/nonseparatist debate noted in chapter one at note 74. Michael S. Baram, Cost-Benefit Analysis: An Inadequate Basis for Health, Safety, and Environmental Regulatory Decisionmaking, 8 Ecology Law Quarterly 473 (1980); Starr and Whipple, Risks of Risk Decisions, 208 Science 1114-19 (June 1980); Harold P. Green, Cost-Risk-Benefit Assessment and the Law: Introduction and Perspective, 45 George Washington Law Review 901 (1977); Amory B. Lovins, Cost-Risk Benefit Assessment in Energy Policy, 45 George Washington Law Review 911 (1977); James S. Fishkin, Tyranny and Legitimacy: A Critique of Political Theories 91-96 (Baltimore: Johns Hopkins University Press 1979); Mark Sagoff, We Have Met the Enemy and He Is Us *or* Conflict and Contradiction in Environmental Law, 12 Environmental Lawyer 283 (1982); Mark Sagoff, At the Shrine of Our Lady of Fatima, or Why Political Questions Are Not All Economic, 23 Arizona Law Review 1283 (1981); Mark Sagoff, Economic Theory and Environmental Law, 79 Michigan Law Review 1393 (1981); Frank I. Michelman, Norms and Normativity in the Economic Theory of Law, 62 Minnesota Law Review 1015 (1978); and Mark Kelman, Cost-Benefit Analysis: An Ethical Critique, 4 Regulation 33 (Jan./Feb. 1981).

2. Shippen Howe, A Survey of Regulatory Treatment of Plant Cancellation Costs, 111 Public Utilities Fortnightly 52 (March 31, 1983); Department of Energy, Energy Information Administration, Nuclear Plant Cancellations: Causes, Costs, and Consequences 33-57 DOE/EIA-0392 (Washington: Department of Energy April 1983); Office of Technology Assessment, Nuclear Power in an Age of Uncertainty 46–57, OTA-E-216 (Washington: February 1984); Richard J. Pierce, Jr., The Regulatory Treatment of Mistakes in Retrospect: Canceled Plants and Excess Capacity, 132 University of Pennsylvania Law Review 497, 517–20 (1984).

3. Office of Consumers' Counsel v. Public Utilities Commission, 67 Ohio St. 2d 153 (1981); Maureen A. Crawfis, New Approach to Allocating Financial Responsibility for Canceled Nuclear Units, 13 Toledo Law Review 1469 (1982); Anne L. Hammerstein, Consumers' Counsel v. Public Utilities Commission: Who Shall Bear the Cost of Abandonment, 11 Capital University Law Review 91 (1981).

4. Some critics say that the concept is a hallucination; Peter Gabel and Duncan Kennedy, Roll Over Beethoven, 36 Stanford Law Review 1 (1984).

5. Mancur Olson, The Logic of Collective Action (Cambridge, Mass.: Harvard University Press 1971), and James Buchanan and Gordon Tullock, The Calculus of Consent: Logical Foundations of Constitutional Democracy (Ann Arbor: University of Michigan Press 1965).

6. John E. Chubb, Interest Groups and the Bureaucracy: The Politics of Energy (Stanford: Stanford University Press 1983). Professor Chubb argues that energy policy is shaped by government and industry interests and that public interest as such is an ineffective competitor.

7. Mark Tushnet, The Constitution of the Bureaucratic State, 86 West Virginia

Law Review 1077 (1984), and Gerald E. Frug, The Ideology of Bureaucracy in American Law, 97 Harvard Law Review 1277, 1296–1317 (1984). Professor Frug describes, in great detail, four bureaucratic models. My immediate remarks and the citation refer to what he calls the formalist model. Richard B. Stewart, The Reformation of American Administrative Law, 88 Harvard Law Review 1669, 1671–88 (1975). Frug makes a telling point when he compares administrative law with corporate law and argues that they share common failings. The nuclear industry and its regulation also have the weaknesses Frug describes, in large part because of the closeness of the joint venture.

8. 10 C.F.R. §1.64 (1984).

9. 10 C.F.R. §55.70 (1984).

10. Campo, The Case against Shoreham, in Report of the New York State Fact Finding Panel on the Shoreham Nuclear Power Facility 38–40 (Dec. 1983); Eric E. Van Loon and Ellen R. Weiss, The State of the Nuclear Industry and the NRC: A Critical View, Testimony before the Nuclear Regulatory Commission (Washington: Union of Concerned Scientists Nov. 17, 1983). The case against the NRC's slighting of safety inspections in favor of licensing was made more fully in chapter four. This neglect is a pervasive part of nuclear regulation and is one of the key lessons of TMI.

11. NRC Office of Inspection and Enforcement, Region III, Letter to Cincinnati Gas & Electric Co., June 10, 1980, Report No. 50-358/80-12 at 3.

12. Diamond, The Heat Is Still Rising on Nuclear Regulators, N.Y. Times, June 29, 1984, p. 24E, col. 4; and Matthew Wald, Panel Concedes Erring in Permit for Atom Plant, N.Y. Times, July 24, 1984, p. 7, col. 1.

13. Government Accountability Project, Request for an Investigation Pursuant to 5 U.S.C. §1206(B)(7), Before the Office of the Special Counsel of the Merit Systems Protection Board (Dec. 10, 1980) at 1. GAP more generally charged the NRC with failure "to perform a thorough and complete investigation of serious allegations" made about the Zimmer facility, at id.

14. 28 U.S.C. §1346(b) (1982).

15. 28 U.S.C. §2680(a) (1982).

16. Dalehite v. United States, 346 U.S. 15, 35 (1953).

17. Peter H. Schuck, Suing Government: Judicial Remedies for Official Wrongs 114 (New Haven: Yale University Press 1983).

18. Blessing v. United States, 447 F. Supp. 1160, 1162 (E.D. Pa. 1978).

19. 346 U.S. 15 (1953).

20. United States v. S.A. Empresa de Viacao Aerea Rio Grandense (Varig Airlines), 465 U.S. 1018, 1052 (1984).

21. 447 F.2d at 1162 (E.D. Pa. 1978).

22. Shuck, note 17 at 115.

23. Gercey v. United States, 409 F. Supp. 946 (D.R.I. 1976).

24. General Public Utilities v. United States, 551 F. Supp. 521 (E.D. Pa. 1982).

25. Id. at 525. The court later wrote at 526: "The NRC's duty to monitor the industry and warn of safety problems does not, however, amount to a *guarantee* of plant safety. . . . Accordingly, we conclude that . . . the NRC has undertaken *some* duty to carefully monitor nuclear experiences and to disseminate appropriate warnings. Moreover, this duty runs, *inter alia*, to the nuclear power industry."

26. 551 F. Supp. at 530–31. The court denied the government's motion to dismiss and certified an interlocutory appeal. The Third Circuit stayed the proceeding until the United States Supreme Court decided the *Varig Airlines* case, note 20, which absolved the government from liability. The GPU case can be distinguished from *Varig Airlines* on several grounds. First, the FAA is a minimal safety inspection agency as compared with the NRC. Second, the record in *Varig Airlines* was spotty. Third, NRC inspectors' duties are more fully detailed.

27. Collins v. United States, 621 F.2d 832 (6th Cir. 1980); Raymer v. United

States, 455 F. Supp. 165 (W.D. Ky. 1978); Clemente v. United States, 567 F.2d 1140 (1st Cir. 1977); Reminga v. United States, 448 F. Supp. 445 (W.D. Mich. 1978); Aretz v. United States, 503 F. Supp. 260 (S.D. Ga. 1977); and Bramer v. United States, 595 F.2d 1141 (9th Cir. 1979).

28. 346 U.S. at 16.

29. 28 U.S.C. §2680(h) (1982).

30. The General Public Utilities v. United States case contains a curious anomaly based on the economic-safety dichotomy. If GPU argues that the government's breach was economic, they may be barred by the misrepresentation exception of the FTCA, 28 U.S.C. §2680(h). The court said the exception was no bar insofar as the cause was safety-related, even though the consequences were business losses. 551 F. Supp. at 529.

31. Schuck, note 17 at 89–90.

32. 416 U.S. 231 (1974).

33. 416 U.S. at 247.

34. Pierce, note 2 at 508–511.

35. 42 U.S.C. §1983 (1982). Schuck, note 17 at 41–57.

36. The cast of characters would also include subcontractors and other vendors. The larger the cast, the more factually complex the case becomes, and causation becomes more problematic; nevertheless, the liability principle remains the same.

37. In November 1979, one prominent litigator wrote, "There are no decided cases setting out broad rules governing litigation about the construction of nuclear facilities. The controversies which have gone to litigation have been settled or are now *sub judice*." Thomas Evans, Construction Litigation Involving Nuclear Facilities, in Thomas Evans (ed.), Nuclear Litigation 203 (New York: Practicing Law Institute 1979).

38. Cincinnati Gas & Elec. v. General Electric Co., Dkt. No. C-1-84-0988 (S.D. Ohio, July 10, 1984) (Complaint).

39. Niagara Mowhawk Power Corp. v. Grauer Tank & Mfg. Co., 470 F. Supp. 1308 (N.D. N.Y. 1979). Utility terminated contract with manufacturer of steel liner for the containment vessel because of dissatisfaction with performance and scheduling delays.

40. Nebraska Public Power District v. General Electric Corp., Dkt. No. Civ. 75-L-142 (D. Neb. June 29, 1979). Court denied GE's motion to dismiss NPPD's claims that GE breached express and implied warranties that it would perform its duties as an expert. Reported in Thomas Evans, Civil Actions by Utilities, in Thomas Evans (ed.), Nuclear Litigation 167 (New York: Practicing Law Institute 1982).

41. The heart of the litigation between owners, builders, and suppliers of nuclear plants is the extent to which contract terms govern. Have the defendants adequately protected themselves with limitations of liability? Have they breached any warranties? Has their conduct gone beyond breach of contractual duty into the area of tort? Damages will be the most contested issue in this type of litigation. Thomas Evans, note 37 at 212, has written:

> Although each case depends on its own facts direct or general damages are usually recoverable in construction cases, while special or consequential damages are not. It is in the manufacturer's interest, however, to permit by contract or usage *some* recovery in order to avoid complete exculpation which is generally not favored as a matter of policy.

42. Rubin v. Dickhoner et al.; Dkt. No. C-1-83-1721 (S.D. Ohio, Feb. 21, 1984) (Complaint).

43. In Susquehanna Valley Alliance v. Three Mile Island Nuclear Reactor, 619 F.2d 231 (3d Cir. 1980) cert. denied 449 U.S. 1096 (1982), the court held that no private right of action exists under the Atomic Energy Act. The AEA is therefore closed as a way of avoiding abandonment costs.

44. Pennsylvania v. General Public Utilities Corp., 710 F.2d 117 (3d Cir. 1983) (case settled).

45. In re TMI Litigation Governmental Entities Claims, 544 F. Supp. 853 (M.D. Pa. 1982).

46. 701 F.2d at 123.

47. Even the Third Circuit in the Pennsylvania v. General Public Utilities case was not particularly hopeful at the end of its opinion. The court wrote, 710 F.2d at 123–24: "Our decision will not preclude the district court from considering any renewed motion for summary judgment or other appropriate disposition of the claims short of a trial, upon establishing of record the material facts."

48. County of Suffolk v. Long Island Lighting Co., 728 F.2d 52 (2d Cir. 1984).

49. The Second Circuit's opinion can be criticized on three grounds.

First, the discussion of federal preemption is not quite accurate. *Pacific Gas & Electric* is clear that the states have power over such things as "the need for additional generating capacity, the type of generating facilities to be licensed, land use, rate making and the like." 103 S.Ct. at 1726. Also, "[M]oreover, Congress has allowed the states to determine—as a matter of economics—whether a nuclear plant vis-à-vis a fossil fuel plant should be built," id. at 1731; and, finally, "The legal reality remains that Congress has left sufficient authority in the states to allow the development of nuclear power to be slowed or stopped for economic reasons," id. at 1132. With language this clear from the United States Supreme Court, the judge's federal preemption analysis fudges the issue when he writes: "In sum, federal preemption encompasses both safety and productivity, and a successful energy policy must strike a delicate balance between the two." Basically, the federal preemption discussion is superfluous other than to argue that there is some interaction between safety and economic issues. No one would deny that.

Second, the court discusses state statutory preemption. The New York PSC generally looks narrowly at ratemaking issues and concentrates on the revenue requirement. The PSC has worked itself into somewhat of a rather cozy, if not dependent, relationship with the utilities. The aim of both the PSC and the utilities is how best to develop a rate formula that will keep the utility afloat. The PSC and utilities keep dumping more costs into the rate base. Although it is true that ratemaking is part of the commission's daily work, it does not follow that the PSC has jurisdiction to look at issues raised in the county's complaint. Such questions as contractor mismanagement, NRC conduct, and some management decisions as bases for damages are generally not examined in the PSC. Therefore, to say that the N.Y. Public Service Law preempts these other claims is wrong.

The final problem with the opinion is that it misreads *Martin v. Julius Dierck Equipment Co.*, to preclude state common-law actions. That case is a choice of law case or a conflicts case and deals with the proper characterization of a products liability cause of action as a tort or a contract. *County of Suffolk* is not a products liability case.

50. Note 42.

51. Pierce, note 2 at 512 n.82: "Indeed, the central thesis of this Article is that many such decisions were imprudent in that they were essentially products of the regulatory incentive to overinvest in capital assets."

52. 67 Ohio St. 2d 153, 423 N.E. 2d 820 (1981).

53. Pierce, note 2 at 511–17; Roger D. Colton, Excess Capacity: Who Gets the Charge from the Power Plant, 34 Hastings Law Journal, 1133, 1143–50 (1983).

54. Charles Phillips, The Regulation of Public Utilities 292–94 (Charlottesville: Michie Bobbs-Merrill Co. 2d ed. 1984).

55. Managerial immunity may not be such a safe haven. In re Columbus and Southern Ohio Electric, 50 PUR 4th 37 (Ohio PUC 1982), the Office of Consumers' Counsel filed a motion with the Ohio Public Utilities Commission against the

management of the Columbus and Southern Ohio Electric Company requesting a management audit in regard to the construction of the Zimmer nuclear power plant. The underlying basis for the motion was to determine whether C&SOE's imprudent policies were the cause of additional costs necessary to complete work on the Zimmer project. The motion was premised on 1) the increased costs of the project; 2) the increase in time necessary to complete construction; 3) the subsequent Nuclear Regulatory Commission investigations and their impact. Although the PUC denied the motion because C&SOE was only one of three utilities involved in the plant's construction and was not controlling the plant's construction, the PUC inferred that the grounds upon which the motion was based could have led to an audit of the utility's management if it had been the controlling partner and that these grounds could have led to the conclusion that management had made an imprudent decision by not abandoning the plant earlier in its construction.

56. The simple test for cost-of-service ratemaking is called the "used and useful" test. Ratepayers will be charged only for a nuclear plant that is on line and operational. The advantage of the test is that it is simple to apply. The test should have something more than administrative ease to commend it. The downside to the "used and useful" formula is that it creates an incentive for managers to continue to pour money into a project, e.g., convert a nuclear plant to coal rather than cancel when the plant is no longer needed.

57. In re Southern California Edison Co., 50 PUR 4th 317 (Cal. PUC 1982).

7. *Transition and Beyond*

1. Compare H. L. A. Hart, The Concept of Law, ch. 2 (Oxford: Clarendon Press 1961) (Hart's is the strongest modern statement of legal positivism as a neutral rules system), with Ronald Dworkin, Taking Rights Seriously, chs. 2 and 3 (Cambridge: Harvard University Press 1978) (law is more than a system of rules; it also consists of policies, principles, and standards).

2. I can identify four models to describe the interaction of law and policy. I have consciously chosen the simplest of the four. The law-follows-policy formula is the dominant mode of interaction in a bureaucracy, which is the level of policy analysis under discussion.

Model I, in which law follows policy, is based on two Legal Realist premises. The first is that law is not value-neutral. The second is that the legal system assists government in the implementation of its policies. G. Edward White, Patterns of American Legal Thought 99–135 (Indianapolis: Bobbs-Merrill Co. 1978), and Bruce Ackerman, Reconstructing American Law, ch. 2 (Cambridge: Harvard University Press 1984).

Model II, in increasing order of complexity, posits that policy follows law. This model is more complex, because it requires a settled definition of law and a partial suspension of belief. This positivist model requires a conception of law as an objective and neutral manifestation of a just order that exists as given, and is then applied to a determinate dispute or policy problem. Policy, then, is crafted to fit the given law. At an operational level, Model II may function. An NRC staffer, for example, may accept the agency's promotional nuclear policy and wish to limit consumer intervention in a licensing hearing. However, the staffer is constrained by the law of due process and must allow meaningful participation. Therefore, policy must follow law. Note that the operational level of analysis is distinct from the more systemic examination of the interaction of law and policy which is the subject of the book.

The third configuration combines the first two. In Model III, law and policy interact in different sequences at different times. Sometimes law affects policy; sometimes policy shapes law. This model adds more detail and raises new questions. Are there identifiable sets of circumstances, or categories, in which one formula governs the other? Further, is there an arrangement or relationship beween categories? If this book were about the interaction of "law" and "policy" as generic topics, I would argue that Model III is a better description of the phenomena. The topic is limited to law and policy in a bureaucracy.

In Model III, the interaction of law and policy is the sum of the manifestations of law and the manifestations of policy. The word *manifestations* should be explained. Law and policy are not concrete "things" in the world. Rather, they are artificial devices used to understand and make sense of the complexities of everyday objective and subjective, individual and collective life. By "interaction of law and policy" I mean nothing more, or less, than that this interaction is a manifestation of the operation of the modern regulatory state.

Model IV can be called the Cybernetic Model. Here law and policy inform, influence, and illuminate each other as information is continuously flowing through two interconnected systems. The systems are not discrete, as is suggested in Model III; instead Model IV is something other (maybe more, maybe less) than the sum of the manifestations of law and policy. In the Cybernetic Model, the interaction of law and policy describes an entirely different manifestation from that of Model III. John Steinbruner, The Cybernetic Theory of Decision: New Dimensions of Political Analysis (Princeton: Princeton University Press 1974).

Model IV is the most attractive. It is also the most problematical. The model is attractive because it is less deterministic than the others. It refuses to order the relationship between law and policy. It is also less reductionist, because it does not demand firm definitions of either law or policy; they can remain fluid. Following these characteristics, the interactions of law and policy in Model IV are more flexible, fluid, and open to possibilities. This last characteristic is the undoing of the model. The possibilities are endless. Law and policy may be arranged hierarchically, cyclically, circularly, statically, dynamically, continuously, discontinuously, or all or none of the above. This last sentence presents the most pressing problem of Model IV: We do not have an adequate vocabulary in which to describe the phenomena that occur within the model.

3. There is a current jurisprudential debate about the public/private nature of law. One way to distinguish categories of substantive law is to dichotomize law into public law and private law. Energy law is a public law subject because it involves numerous interests, affects several groups, and is primarily prospective. Contract law can be described as a system of private law that is geared to ordering arrangements between private individuals and settling past disputes. The dichotomy is not wholly successful. Private law decision making also has prospective applications. When litigating even private disputes, litigants can appeal to a judge's sense of public policy by asserting public collectivist rather than private individualist arguments. In this fashion, private law rules incorporate public policy rationales. There is no set of meta-rules to guide decision makers about when it is proper to accept individualist arguments over collectivist arguments. Gerald Frug, The City as Legal Concept, 93 Harvard Law Review 1059, 1128–49 (1980), and Robert Mnookin, The Public/ Private Dichotomy: Political Disagreement and Academic Repudiation, 130 University of Pennsylvania Law Review 1429 (1982).

4. Charles Reich, The Law of the Planned Society, 75 Yale Law Journal 1227 (1966).

5. Charles Lindblom, Politics and Markets (New York: Basic Books 1977) (the book is a thorough analysis of various social configurations, which differ in the extent to which governments replace markets).

6. Hart, note 1 at 1:

> Few questions concerning human society have been asked with such persistence and answered by serious thinkers in so many diverse, strange, and even paradoxical ways as the question 'What is law?' Even if we confine our attention to the legal

> theory of the last 150 years and neglect classical and medieval speculation about the 'nature' of law, we shall find a situation not paralleled in any other subject systematically studied as a separate academic discipline.

7. Thomas Dye, Policy Analysis 1 (University: University of Alabama Press 1976).

8. Ackerman, note 2, chs. 1 and 2, and Bruce Ackerman and Warren Hassler, Clean Coal/Dirty Air, ch. 1 (New Haven: Yale University Press 1981).

9. Armory Lovins, Soft Energy Paths (San Francisco: Friends of the Earth International 1977).

10. Gerald Frug, The Ideology of Bureaucracy in American Law, 97 Harvard Law Review 1276 (1984).

11. "Rules jurisprudence" is a rich field of law which ranges from Hart's positivism, note 1, to Dworkin's legal liberalism, note 1, to the more politicized Marxian version of the Critical Legal Studies Movement, Duncan Kennedy, Form and Substance in Private Law Adjudication, 89 Harvard Law Review 1685, 1687–1713 (1976); and David Kairys, Introduction and ch. 1 in The Politics of Law (David Kairys ed.) (New York: Pantheon Books 1982).

12. Dye, note 7; Aaron Wildavsky, Speaking Truth to Power: The Art and Craft of Policy Analysis 1–19 (Boston: Little, Brown 1979); Colin Diver, Policymaking Paradigms in Administrative Law, 95 Harvard Law Review 393, 396–401 (1981); and James DeLong, Informal Rulemaking and the Integration of Law and Policy, 65 Virginia Law Review 257, 329–54 (1979).

There are as many conceptual levels of policy analysis as there are definitions. Policy can be analyzed at an operational level or at departmental, agency, or branch-of-government levels. One can also pose the question, What is the policy of law? Clearly, the crucial step is articulating the issue. The policy questions involved here are: How do nuclear power law and policy interact? And, what are the broader implications for modern bureaucracy?

13. Lindblom, note 5.

14. Precisely how principles of equality and fairness apply to modern society comes within the purview of political theories. John Rawls, A Theory of Justice (Cambridge: Harvard University Press 1971); Robert Nozick, Anarchy, State, and Utopia (New York: Basic Books 1974); and Bruce Ackerman, Social Justice in the Liberal State (New Haven: Yale University Press 1980). More narrowly, the idea of government regulation is based on the concept of market failure, Stephen Breyer, Analyzing Regulatory Failure: Mismatches, Less Restrictive Alternatives, and Reform, 92 Harvard Law Review 547, 552–60 (1979).

15. Ackerman, note 2.

16. Breyer, note 14 at 560–78; Robert Litan and William Nordhaus, Reforming Federal Regulation, ch. 4 (New Haven: Yale University Press 1983).

17. There is no shortage of reform proposals. General (or generic) reforms are included in: Stephen Breyer, Regulation and Its Reform (Cambridge: Harvard University Press 1982), and Robert E. Litan and William D. Norhaus, Reforming Federal Regulation (New Haven: Yale University Press 1983). Reforms for the electric industry are contained in Peter Navarro, The Dimming of America: The Real Costs of Electric Utility Regulatory Failure (Cambridge: Ballinger Publishing Co. 1985), and Douglas D. Anderson, Regulatory Politics and Electric Utilities: A Case Study in Political Economy (Boston: Auburn House Publishing Co. 1981). Nuclear reforms appear in Union of Concerned Scientists, Safety Second: A Critical Evaluation of the NRCs First Decade (Washington: Union of Concerned Scientists February 1985), and Report of the President's Commission on the Accident at Three Mile Island, The Need for Change: The Legacy of TMI (Washington: October 1979).

18. Joseph P. Tomain, Constructing a Way Out of the Liberal Predicament, 1985 American Bar Foundation Research Journal 345.

INDEX